Imaginary Abacus - Instruction book

A mind math step-by-step guide to addition and subtraction using an imaginary Japanese abacus (Soroban).

Author:
Paul Green

First published 2017
Edition 2

Copyright @ 2017, Paul Green. All rights reserved.

No part of this publication may be reproduced in any material form (including photocopying or storing in any medium by electronic means and whether or not transiently or incidentally to some other use of this publication) without the written permission of the copyright holder.

ISBN-13: 978-1978043497
ISBN-10: 197804349X

Important

To complete the course in this book you will also need:

1) The accompanying workbook:
   'Imaginary Abacus - Workbook' (ISBN: 978-1542357227).

2) A 13 column (or more) Japanese abacus.

Preface

The Japanese abacus has been used as a calculation tool for generations and can still be seen in use in Japan today. Children are still taught to use this instrument in schools today. It is widely available, cheap to buy and fun to use.

This book will teach you the skills required to use the actual abacus effectively, then how to use an imaginary abacus (also known as a mental abacus).

Once learnt and practised these skills will stay with you throughout your life. A useful and impressive skill that would be an asset for anyone.

CONTENTS

How to follow this training course 6

Workbook format 7

Introduction 8

Parts of the Japanese abacus 9

Putting your numbers in the correct column 10

What amount is each column worth? 11

Abacus basics 12

Register a number with 2 digits on the abacus 20

Register multi-digit numbers on the abacus 22

Moving the beads 24

Addition 27

Imaginary abacus 33

Not enough beads in the column for the addition 36

Addition of 3 or more digit numbers 40

Skipped columns when adding 46

Addition of 3 or more numbers 48

Subtraction 51

Subtracting numbers that have different amounts of digits 52

Not enough beads in the column for the subtraction 58

Skipped columns when subtracting 65

Subtraction of 3 or more numbers 68

Addition and Subtraction together 71

Using the reusable workbook pages 75

Blank sheets for reusable workbook page answers 78

How to follow this training course

THE BEST WAY TO PROCEED

① Read through the instructions in this book, at your own pace, until you see this note:

> Time to use the **workbook**! Go to workbook **page 10**.

② Open your workbook at the page stated and follow the work given until you see this note in your workbook:

> Time to use the **instruction book**! Go to instruction book **page 12**.

③ Go back to this instruction book and continue in this way.

④ Keep a track of your progress by ticking the appropriate box next to the instruction work and your workbook work (see pages 4 & 5 in your workbook).

INSTRUCTION BOOK	Part 1	✓	WORKBOOK	Page 8	✓
			WORKBOOK	Page 9	✓
			WORKBOOK	Page 10	✓
			WORKBOOK	Page 11	✓
			WORKBOOK	Page 12	✓
			WORKBOOK	Page 13	✓
INSTRUCTION BOOK	Part 2		WORKBOOK	Page 14	
			WORKBOOK	Page 15	

Workbook format

Part 1

THE WORKBOOK WILL HAVE THREE TYPES OF WORK

1 Work to test the use of the actual abacus:

Find the addition of the following, use your abacus:

				Abacus
1) 12 + 15		2) 14 + 21		
3) 16 + 25		4) 22 + 31		
5) 58 + 66		6) 55 + 89		

2 Work for skill building, use a pencil. Example, drawing the beads on the abacus:

Draw the beads on the empty abacus drawings:

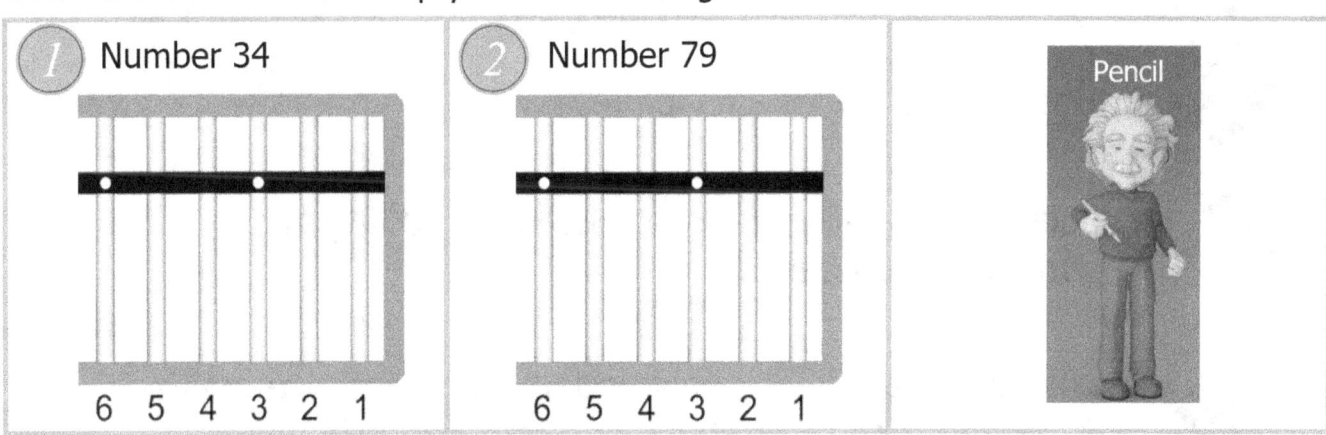

3 Work for using an imaginary abacus:

Find the addition of the following, use an imaginary abacus:

				Imagine
1) 10 + 31		2) 12 + 24		
3) 18 + 19		4) 11 + 25		
5) 68 + 89		6) 80 + 91		

Introduction

Nice to know

 The Japanese abacus is also called the Soroban.

 This is abacus written in Japanese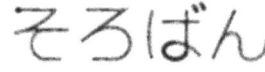

 The Japanese abacus is mostly used for adding and subtracting numbers.

 The Japanese abacus has a wooden frame and five beads per column, one bead above and four beads below.

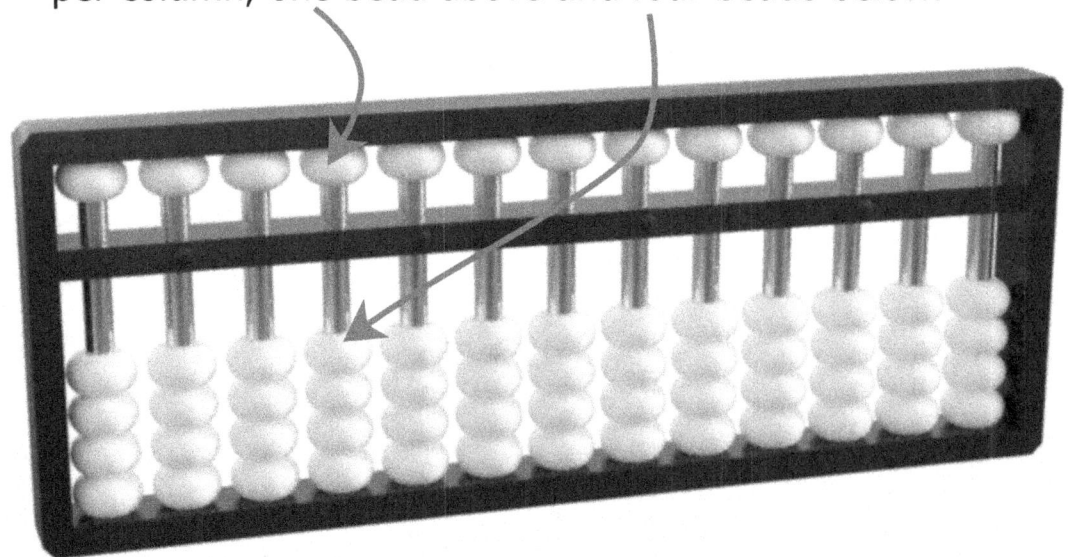

Parts of the Japanese abacus

① A wooden frame.

② A beam, to push the beads up against and away from.

③ Dots on the beam.

④ Rods, to slide the beads up and down on.

⑤ 1 bead above the beam.

⑥ 4 beads below the beam.

⑦ A Column is one rod and the 5 beads on that rod. There are 13 columns on this abacus.

⑧ Lower deck. All beads that are below the beam are in the lower deck.

⑨ Upper deck. All beads that are above the beam are in the upper deck.

Tip Abaci have different amounts of rods. Usually 13 rods but some have less and some have more.

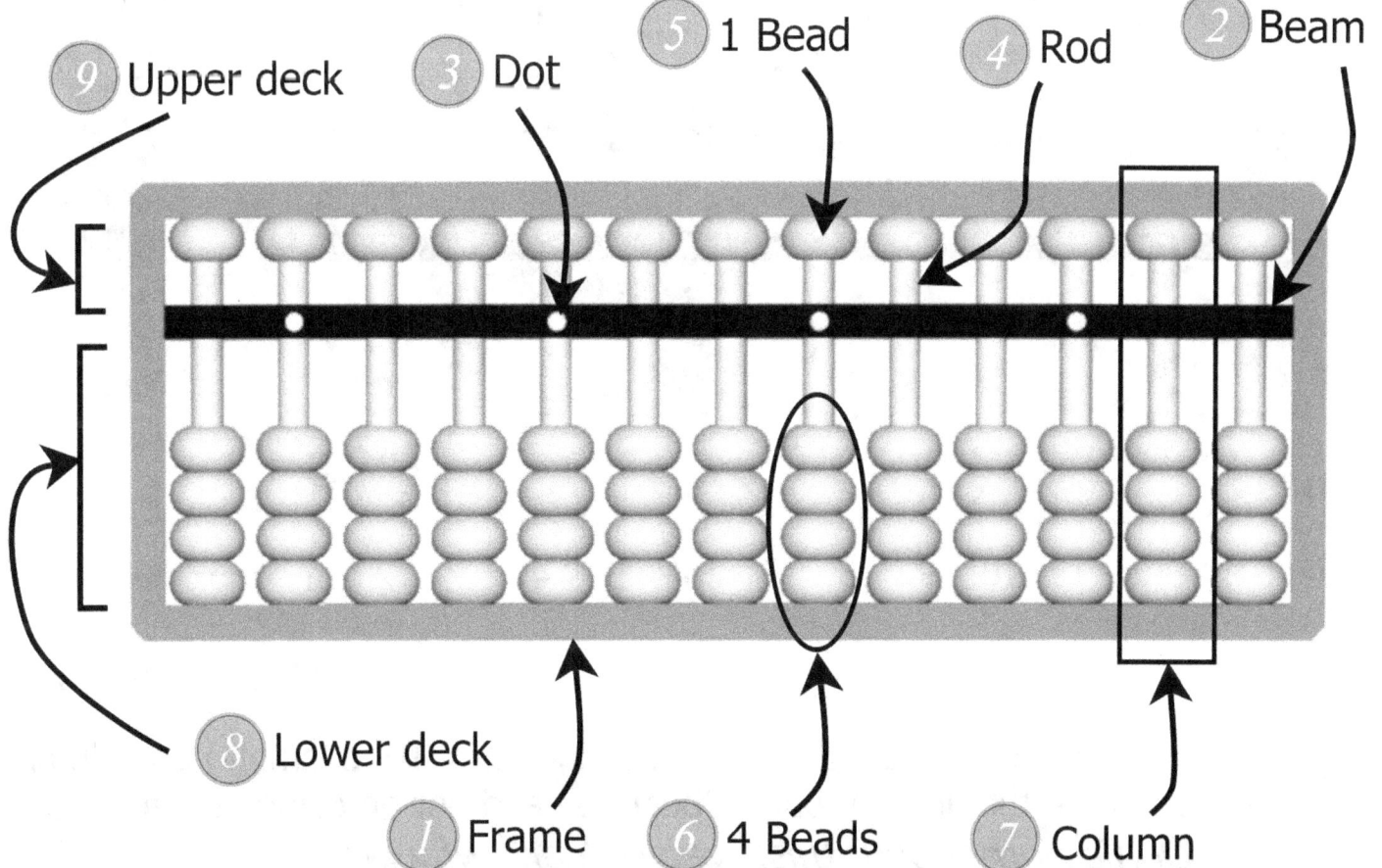

Putting your numbers in the correct column

We need to know which abacus column to use to place each digit.

Let's look at the following 4 digit number **4213**

- The number 3 is on the 'Ones' column
- The number 1 is on the 'Tens' column
- The number 2 is on the 'Hundreds' column
- The number 4 is on the 'Thousands' column

This is how it would look on the abacus.

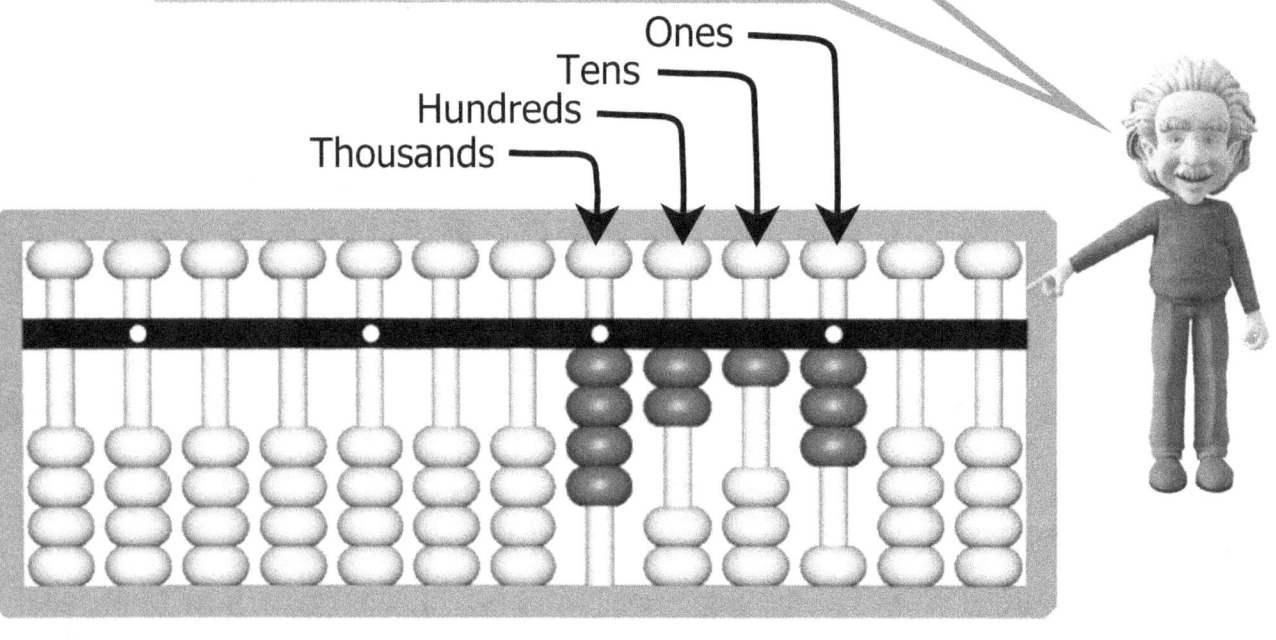

Look how we start with the 'Ones' digit (3 in this example) on column 3 (where the first dot on the beam is) and not on columns 1 and 2. *Don't worry about this, we will learn why later.*

What amount is each column worth?

The picture below shows the values of each column on the abacus.

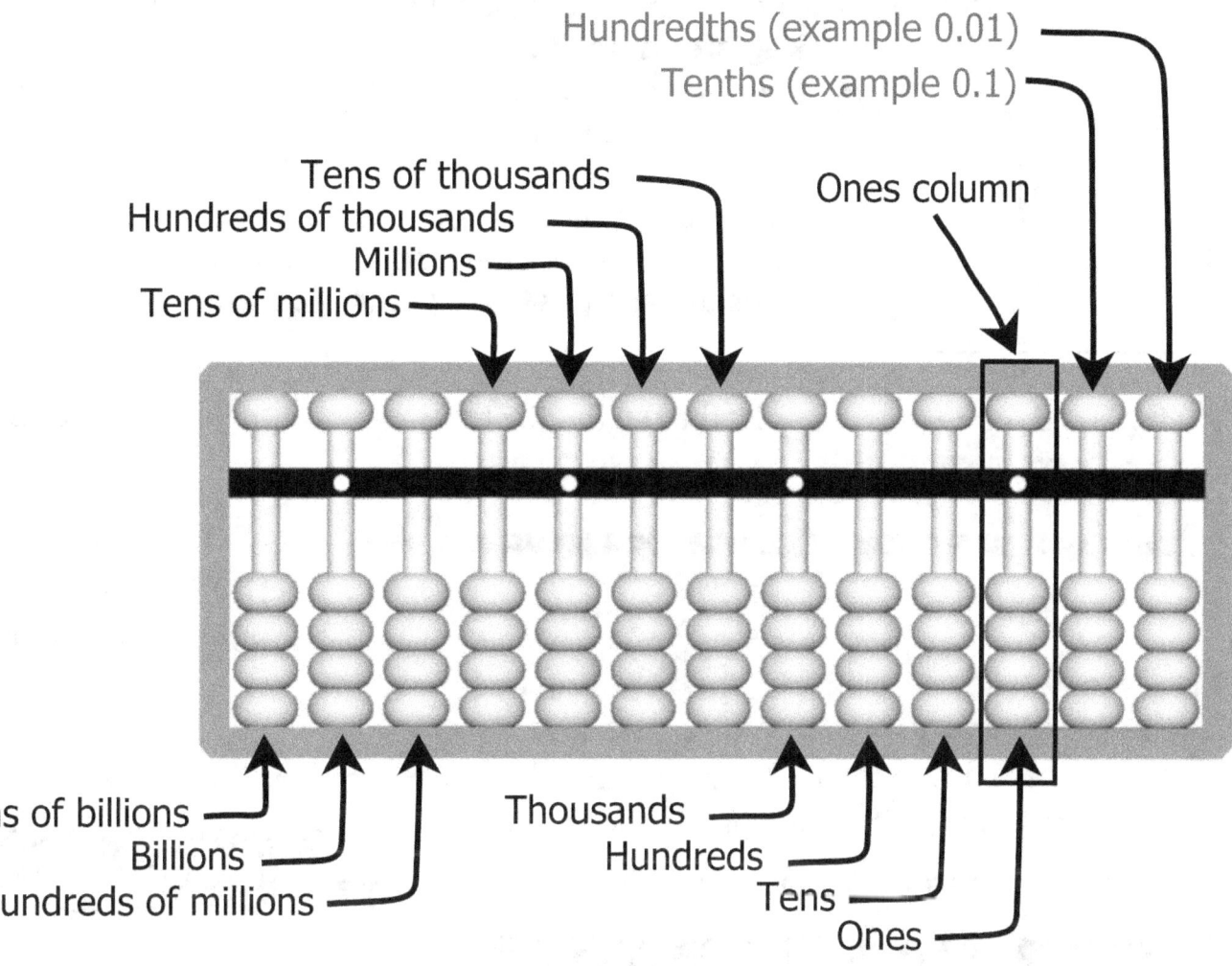

Tip: Notice how each column value keeps getting TEN times BIGGER on the left of the ones column and keeps getting TEN times smaller on the right of the ones column.

You don't need to remember all this to use the abacus. Just remember where the ones column is to get started.

Abacus basics

HOW TO MOVE BEADS TO MAKE A NUMBER ON THE ABACUS

Important!

- When we move a bead towards the beam this is called **'Register a bead'**.
- When we move a bead away from the beam this is called **'Unregister a bead'**.
- The bead above the beam is worth **5**.
- The beads below the beam are worth **1**.
- We read the result on the abacus by looking only at the beads that are **pushed against the beam.**

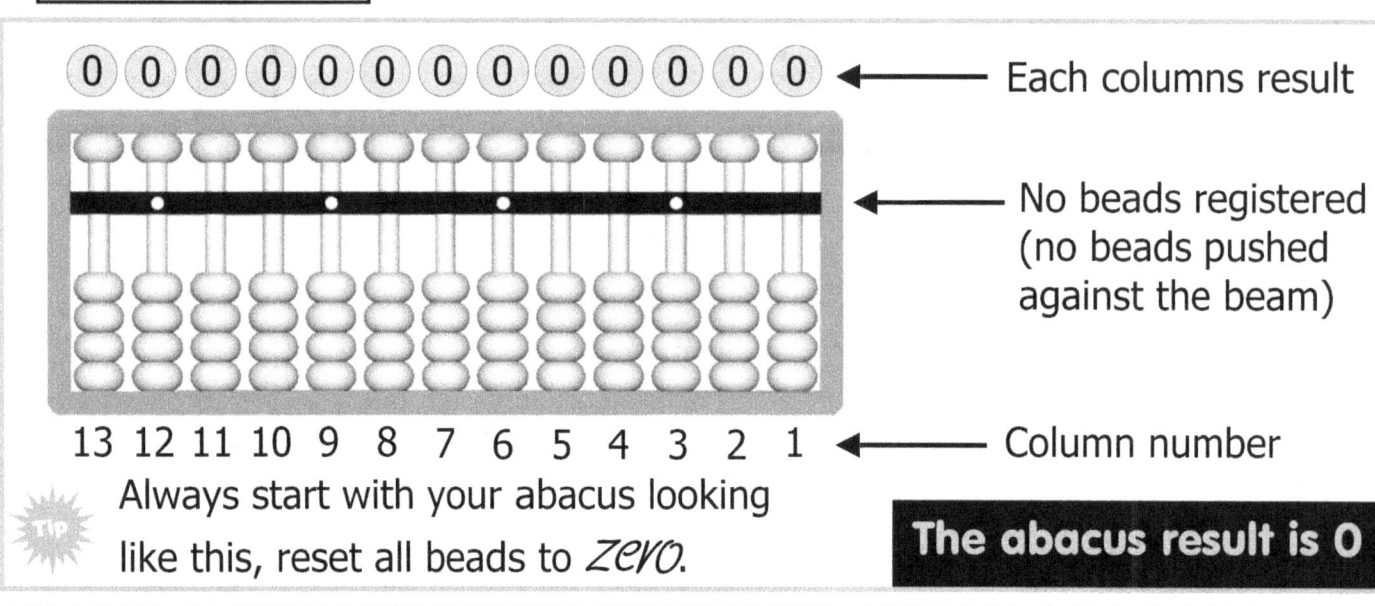

← Each columns result

← No beads registered (no beads pushed against the beam)

← Column number

Always start with your abacus looking like this, reset all beads to *zero*.

The abacus result is 0

Let's put the number 1 on the abacus

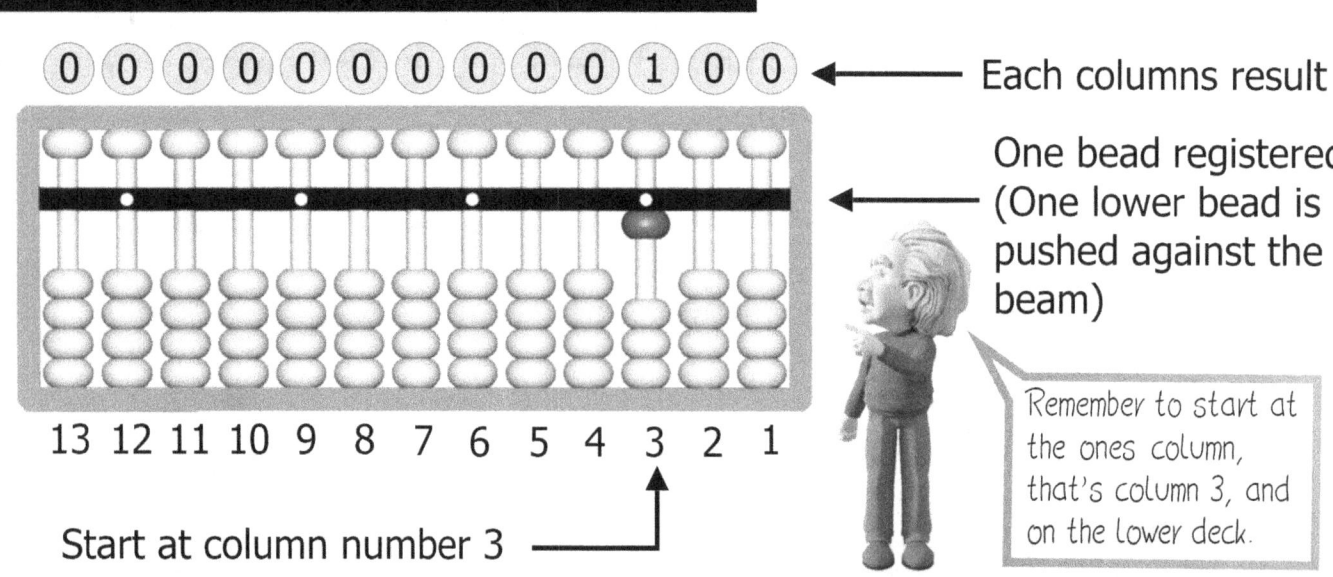

← Each columns result

← One bead registered (One lower bead is pushed against the beam)

Remember to start at the ones column, that's column 3, and on the lower deck.

Start at column number 3

The abacus result is 1

Let's put the number 5 on the abacus

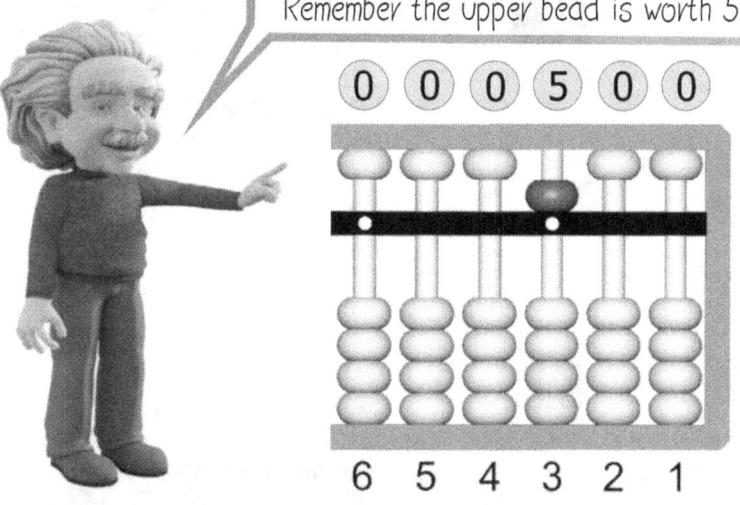

Remember the upper bead is worth 5.

Each columns result

One bead registered (One upper bead is pushed against the beam)

The abacus result is 5

Let's put the number 6 on the abacus

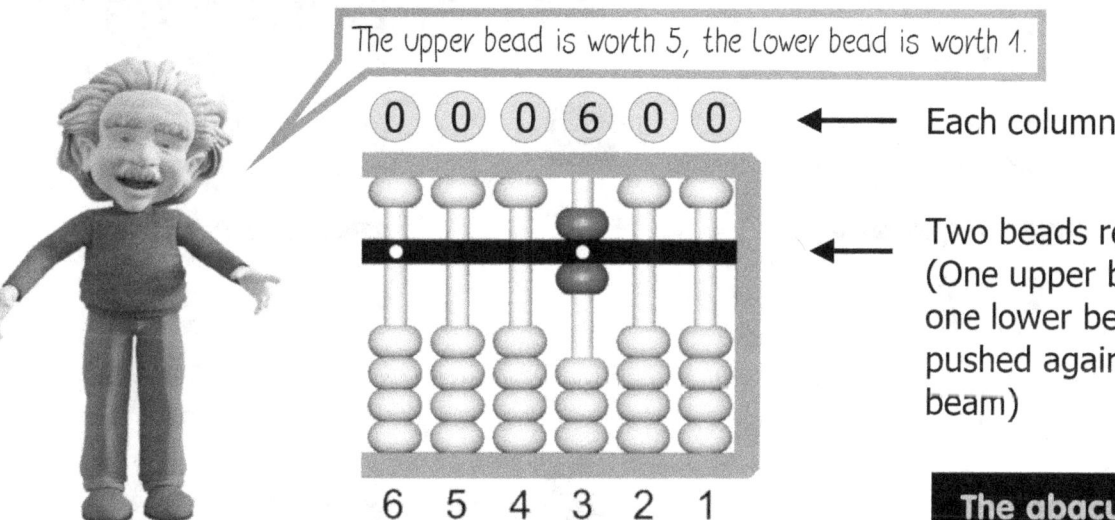

The upper bead is worth 5, the lower bead is worth 1.

Each columns result

Two beads registered (One upper bead and one lower bead are pushed against the beam)

The abacus result is 6

Let's put the number 8 on the abacus

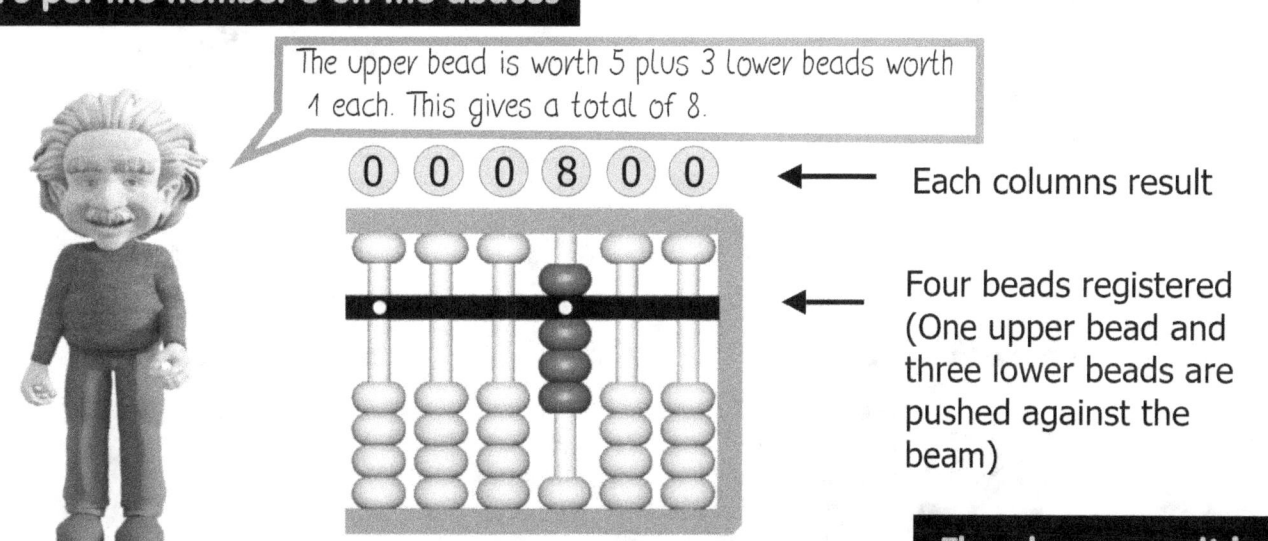

The upper bead is worth 5 plus 3 lower beads worth 1 each. This gives a total of 8.

Each columns result

Four beads registered (One upper bead and three lower beads are pushed against the beam)

The abacus result is 8

Here are the single digit numbers on the abacus

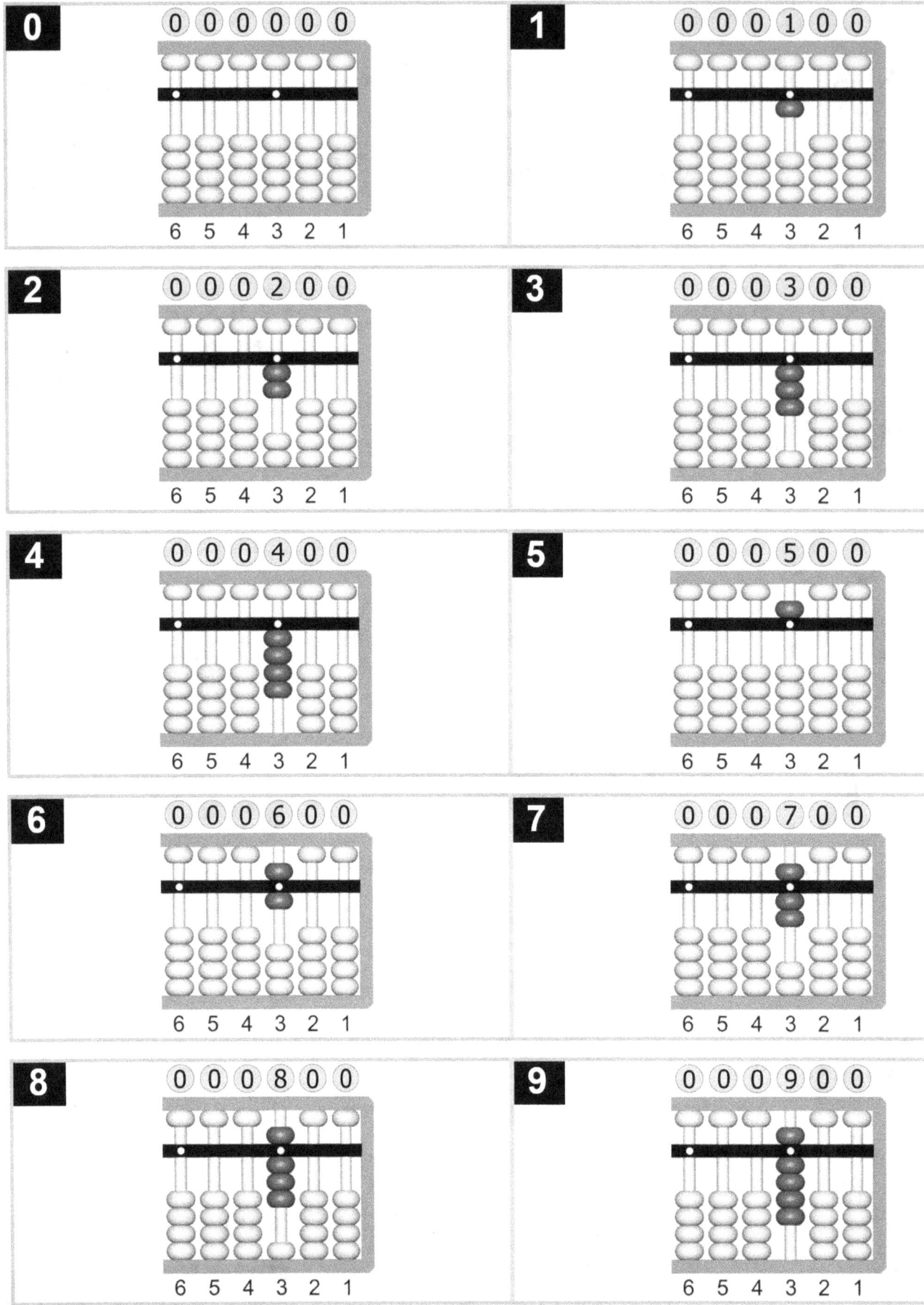

Here are the double digit numbers on the abacus

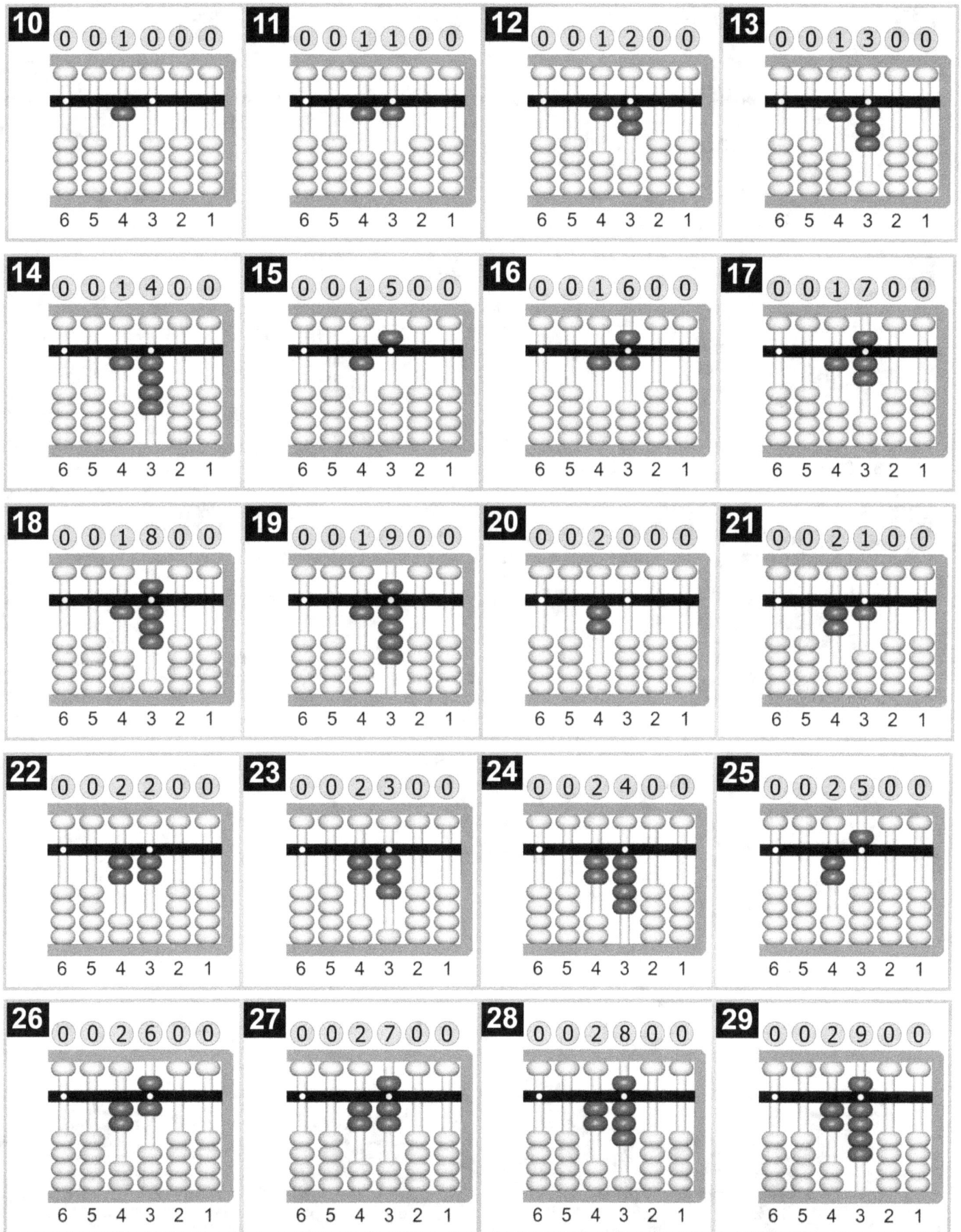

Here are the double digit numbers on the abacus

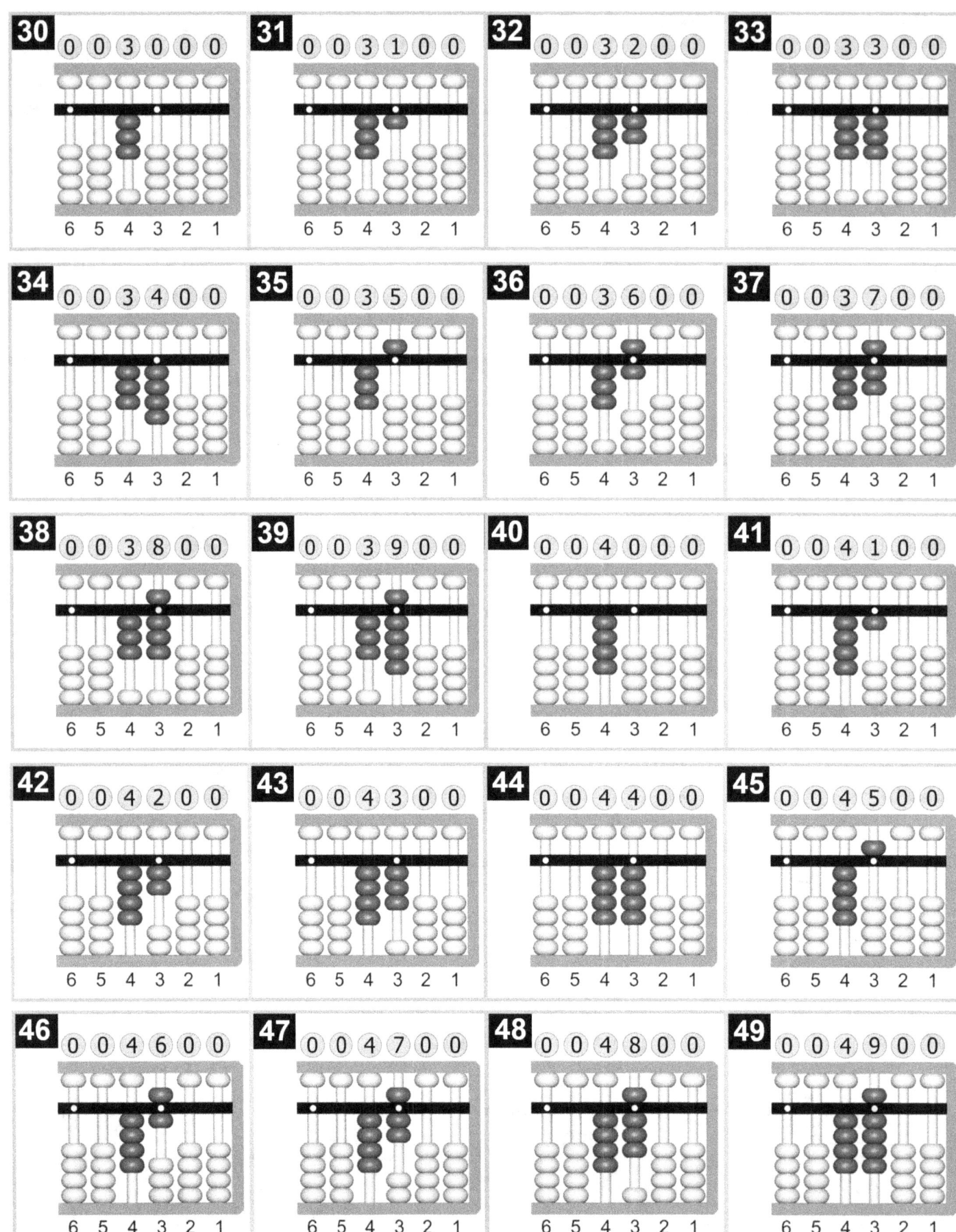

Here are the double digit numbers on the abacus

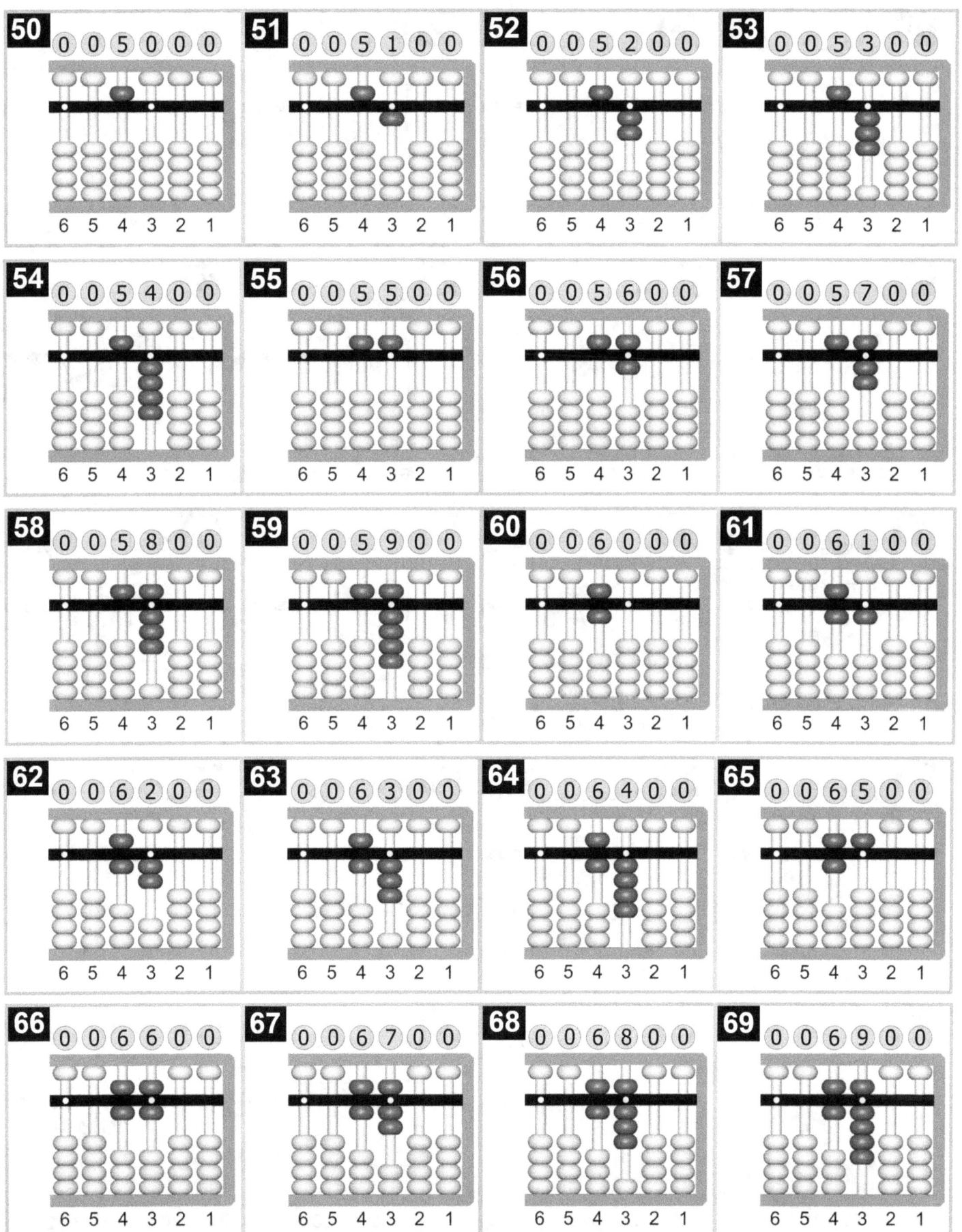

Here are the double digit numbers on the abacus

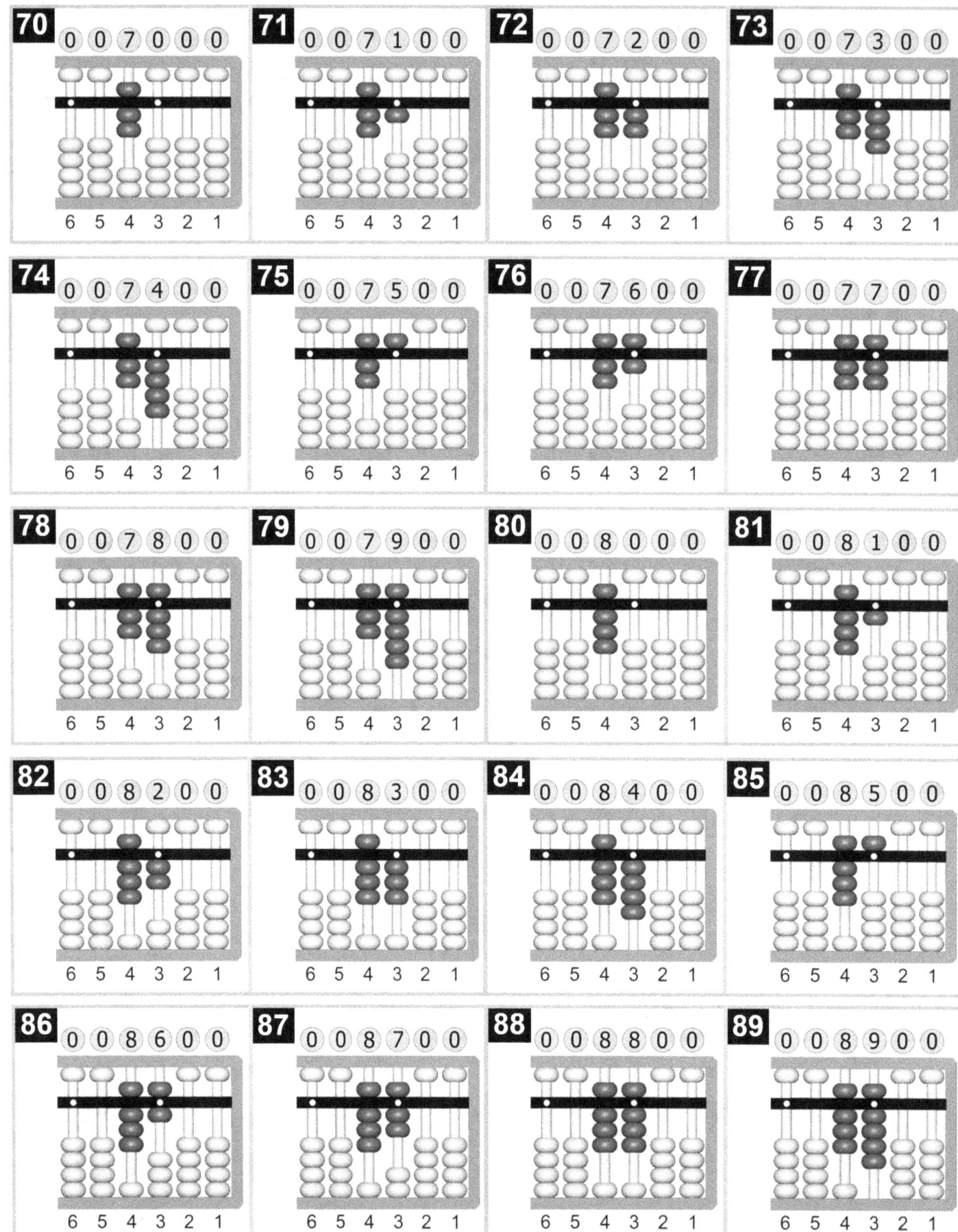

Here are the double digit numbers on the abacus

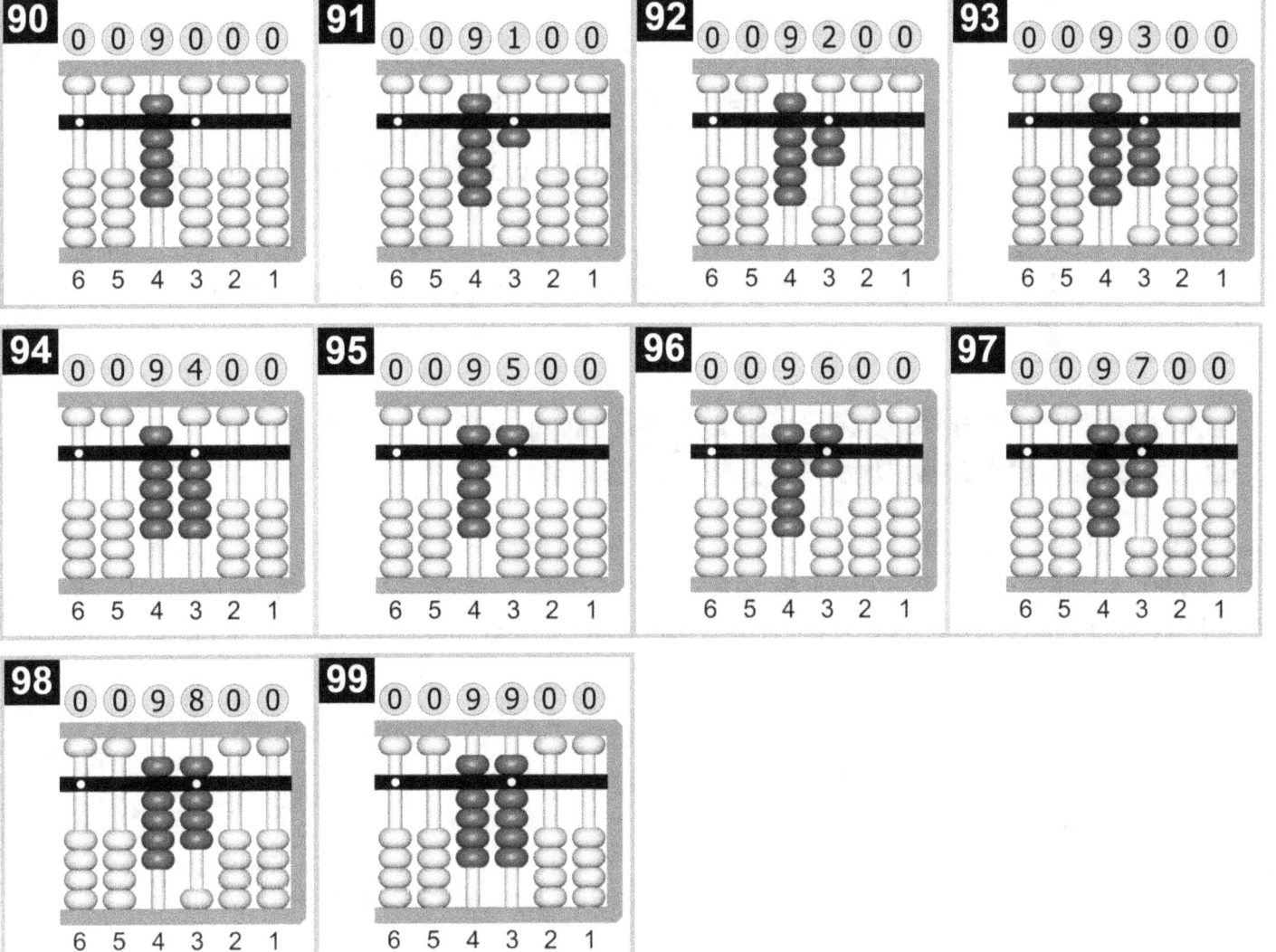

Time to use the **workbook!** Go to workbook **page 8**

Register a number with 2 digits on the abacus — Part 2

Important!

- A **digit** is a symbol used to show a number. Example, **6** is one digit and is made up of one number.
- A digit is any number from **0** to **9**.
- The number **78** has two digits, **7** and **8**. There are two digits that make up the number 78.
- There are two places in the number 78, the ones place holds the number 8 and the tens place holds the number 7.

Let's put the number 15 on the abacus

When we put a number on the abacus or '**register**' a number by pushing the beads towards the beam, we start with the leftmost digit first (in this example the 1 of the 15) then move to the right to register the other numbers.

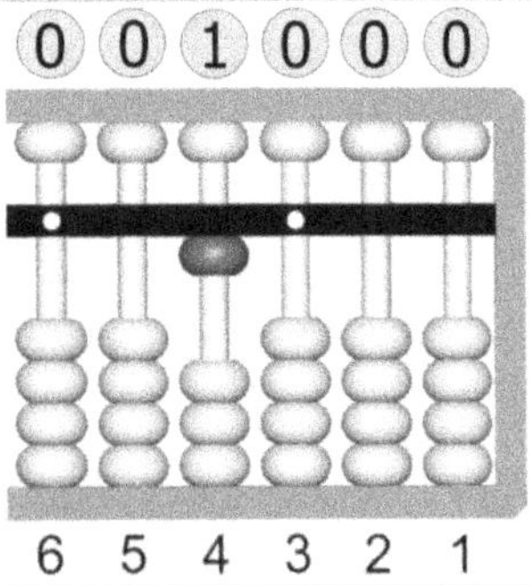

0 0 1 0 0 0

6 5 4 3 2 1

First register the '1' digit (leftmost) in the 'tens column', column 4. → 15

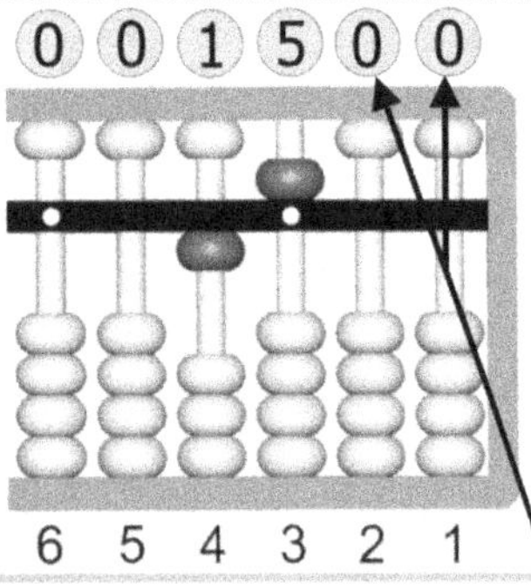

0 0 1 5 0 0

6 5 4 3 2 1

Second register the '5' digit in the 'ones column', column 3. → 15

The abacus result is 15

The two zeros after the number 15 are for decimal numbers, tenths (example the digit 3 in 0.3) for column 2 and hundredths in column 1 (example the digit 6 in 0.46).
We can **ignore** them as we are using whole numbers (also called counting numbers 1, 2, 3, 4....).

Here are some two digit numbers on the abacus

35

⓪ ⓪ ③ ⑤ ⓪ ⓪

6 5 4 3 2 1

- We will put 35 on the abacus
- 35 has 2 digits, so use 2 columns
- Column 4, register 3 lower beads (this is for the 3 of the 35)
- Column 3, register 1 upper bead (this is for the 5 of the 35)

The abacus result is 35

24

⓪ ⓪ ② ④ ⓪ ⓪

6 5 4 3 2 1

- We will put 24 on the abacus
- 24 has 2 digits, so use 2 columns
- Column 4, register 2 lower beads (this is for the 2 of the 24)
- Column 3, register 4 lower beads (this is for the 4 of the 24)

The abacus result is 24

95

⓪ ⓪ ⑨ ⑤ ⓪ ⓪

6 5 4 3 2 1

- We will put 95 on the abacus
- 95 has 2 digits, so use 2 columns
- Column 4, register 1 upper bead and 4 lower beads (this is for the 9 of the 95) Total on this column is 5+4=9
- Column 3, register 1 upper bead (this is for the 5 of the 95)

The abacus result is 95

58

⓪ ⓪ ⑤ ⑧ ⓪ ⓪

6 5 4 3 2 1

- We will put 58 on the abacus
- 58 has 2 digits, so use 2 columns
- Column 4, register 1 upper bead (this is for the 5 of the 58)
- Column 3, register 1 upper bead and 3 lower beads (this is for the 8 of the 58) Total on this column is 5+3=8

The abacus result is 58

87

⓪ ⓪ ⑧ ⑦ ⓪ ⓪

6 5 4 3 2 1

- We will put 87 on the abacus
- 87 has 2 digits, so use 2 columns
- Column 4, register 1 upper bead and 3 lower beads (this is for the 8 of the 87). Total on this column is 5+3=8
- Column 3, register 1 upper bead and 2 lower beads (this is for the 7 of the 87). Total on this column is 5+2=7

The abacus result is 87

Register multi-digit numbers on the abacus

A multi-digit number is any number that has more than one digit.

We have already looked at some multi-digit numbers on the previous page (two digit numbers).

Now we will look at larger numbers. Let's start with a 5 digit number.

23456

First register the '2' digit (leftmost) in the 7th column.

Why the 7th column? Because the number has 5 digits and we are not using the first 2 columns.

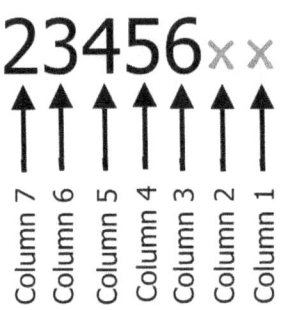

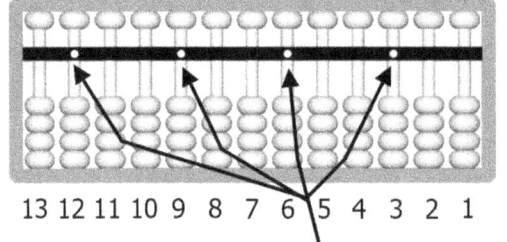

To find the column number quickly, look at the dots. The dots are every 3rd column, 3, 6, 9, and 12.

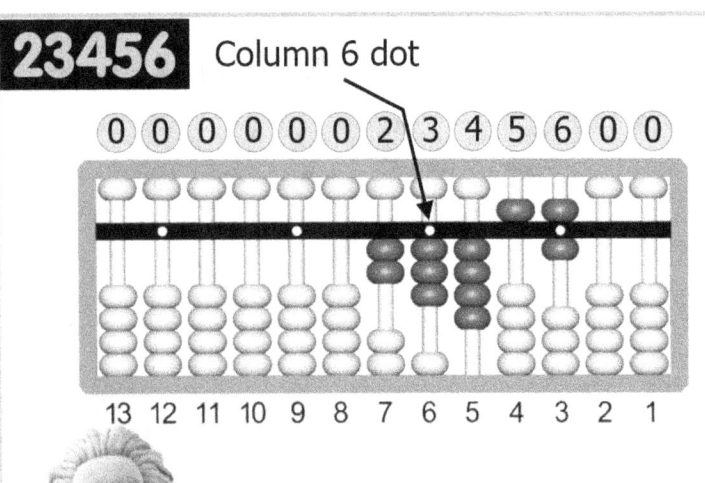

Column 6 dot

- We will put 23456 on the abacus
- 23456 has 5 digits, so use 5 columns (start on column 7)

- Column 7, register 2 lower beads
- Column 6, register 3 lower beads
- Column 5, register 4 lower beads
- Column 4, register 1 upper bead
- Column 3, register 1 upper bead and 1 lower bead
  (total on this column is 5+1=6)

The abacus result is 23456

Things to remember before we move on:
- Don't use columns 1 and 2 (keep those for decimal numbers)
- The total digits of the number plus 2 = the column where we start to register our number
- The dots help us find the column number

Here are some multi-digit numbers on the abacus

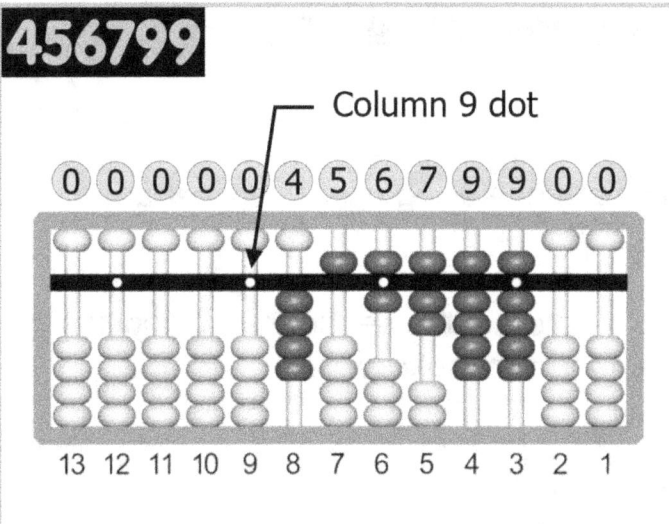

- We will put 456799 on the abacus
- 456799 has 6 digits, so use 6 columns (start on column 8)
- Column 8, register 4 lower beads
- Column 7, register 1 upper bead
- Column 6, register 1 upper bead and 1 lower bead
- Column 5, register 1 upper bead and 2 lower beads
- Column 4, register 1 upper bead and 4 lower beads
- Column 3, register 1 upper bead and 4 lower beads

The abacus result is 456799

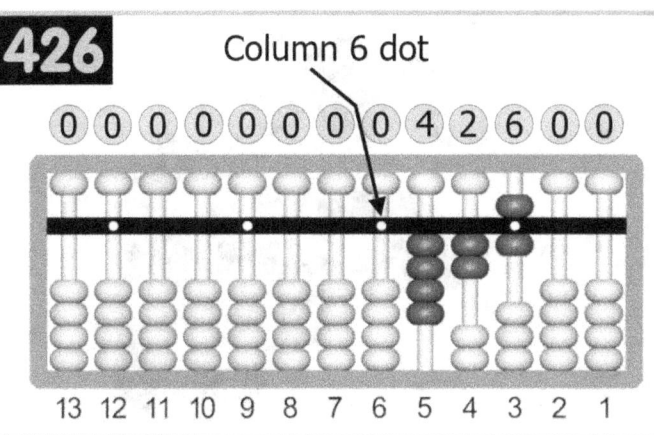

- We will put 426 on the abacus
- 426 has 3 digits, so use 3 columns (start on column 5)
- Column 5, register 4 lower beads
- Column 4, register 2 lower beads
- Column 3, register 1 upper bead and 1 lower bead

The abacus result is 426

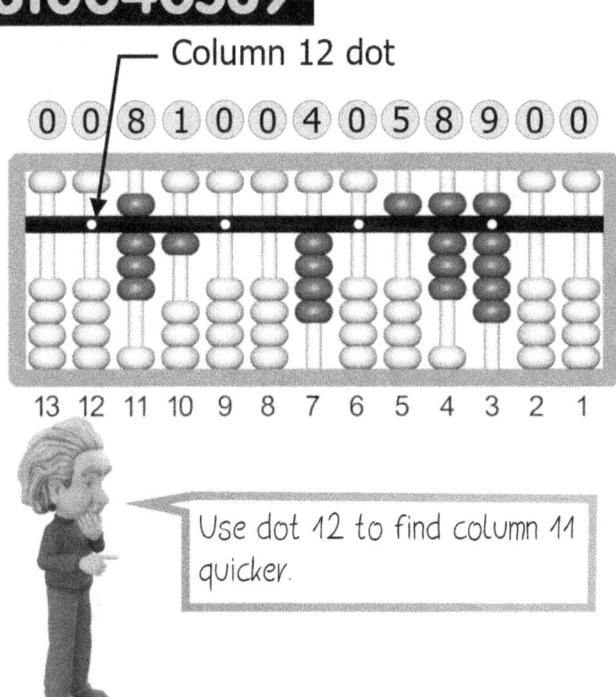

- We will put 810040589 on the abacus
- 810040589 has 9 digits, so use 9 columns (start on column 11)
- Column 11, register 1 upper bead and 3 lower beads
- Column 10, register 1 lower bead
- Column 9, do nothing
- Column 8, do nothing
- Column 7, register 4 lower beads
- Column 6, do nothing
- Column 5, register 1 upper bead
- Column 4, register 1 upper bead and 3 lower beads
- Column 3, register 1 upper bead and 4 lower beads

Use dot 12 to find column 11 quicker.

The abacus result is 810040589

Moving the beads
What fingers do you use to move the beads?

The pictures on pages 25 & 26 will show you what fingers you use to register and unregister the beads.

We will use the thumb and index finger. Some people like to use the thumb, index and middle finger but we will keep it simple for the imaginary abacus.

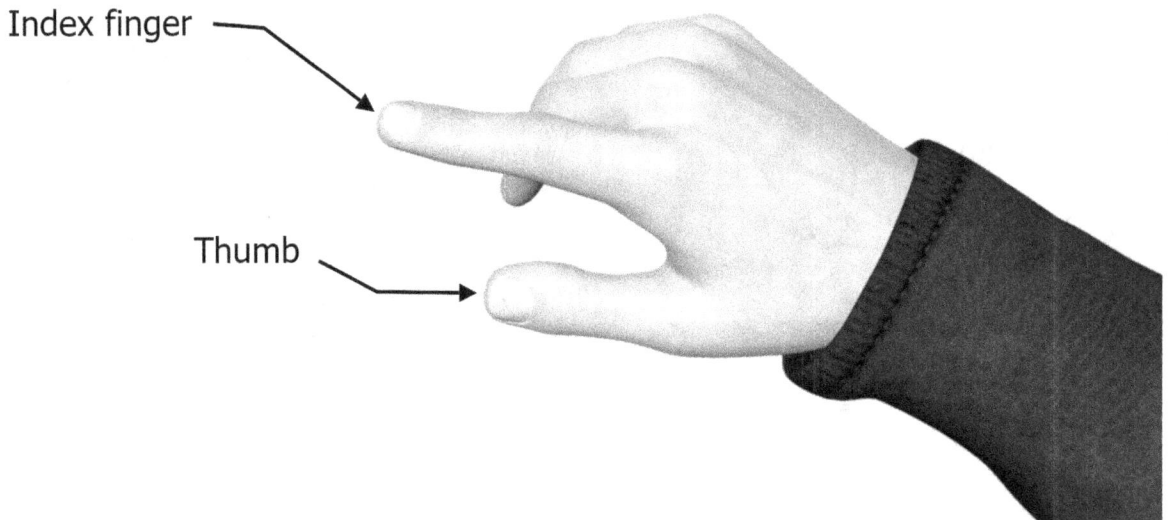

- **Thumb**
  Used to register the **LOWER** beads (to push them towards the beam).

- **Index finger**
  Used to register (towards the beam) and unregister (away from the beam) **ALL other** beads.

- **Bead order**
  When registering and unregistering a number, always start with the highest value column first, then work towards the lowest value column.
  For example, when registering the number 23 start by registering the 2 (20) and then the 3 (3).

On the next pages there are pictures which show the finger movements for registering and unregistering.

Using the thumb

Register 1 lower bead	Register 2 lower beads
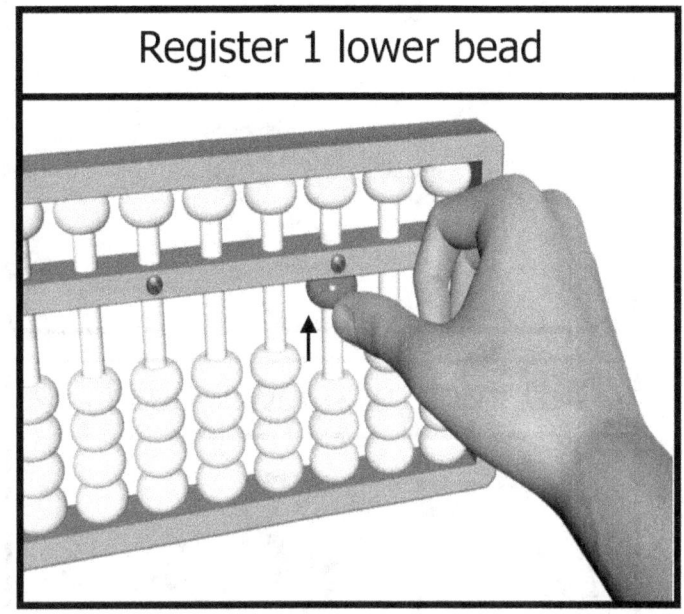	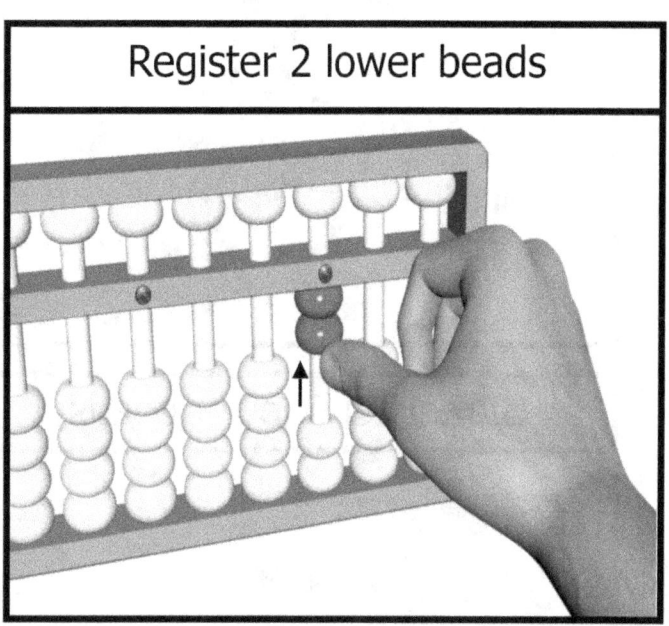
Register 3 lower beads	Register 4 lower beads
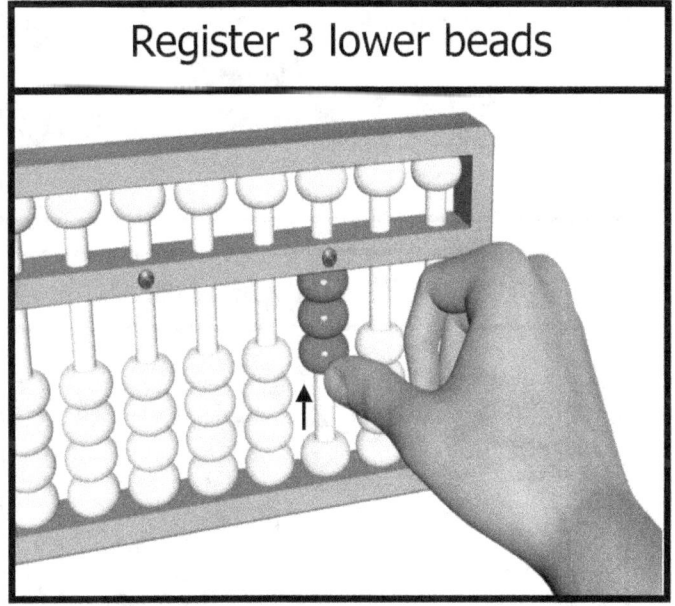	

Using the index finger

Unregister 1 lower bead	Unregister 2 lower beads
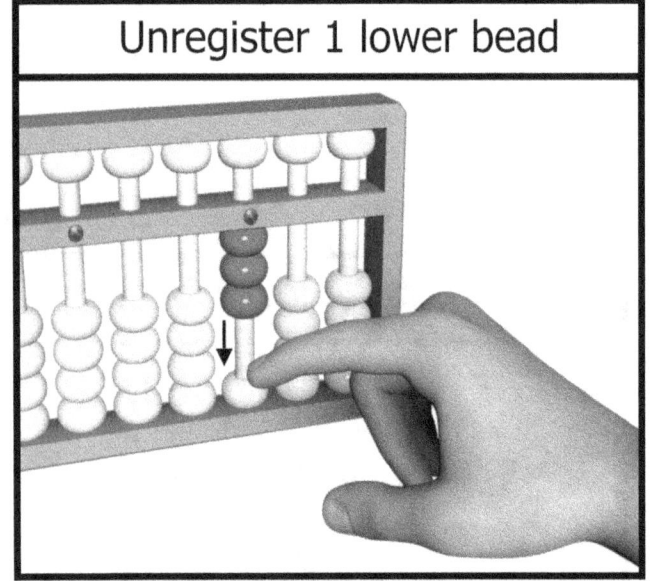	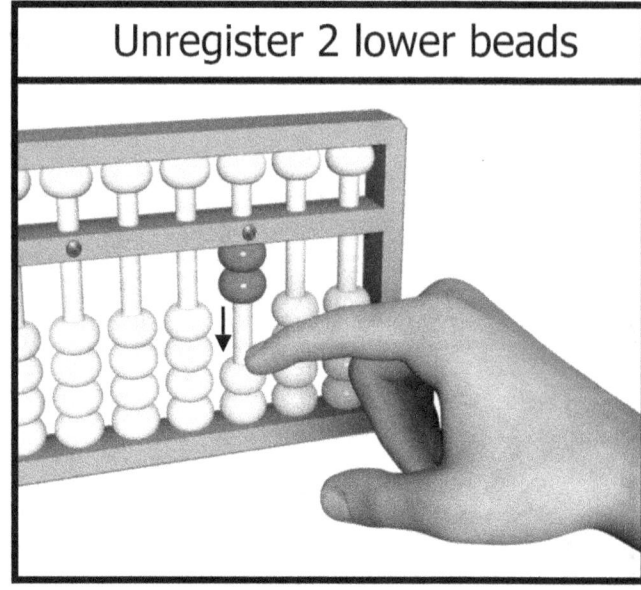
Unregister 3 lower beads	Unregister 4 lower beads
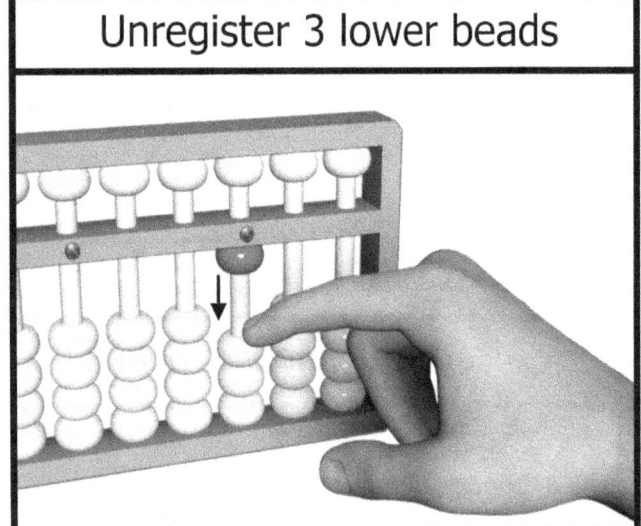	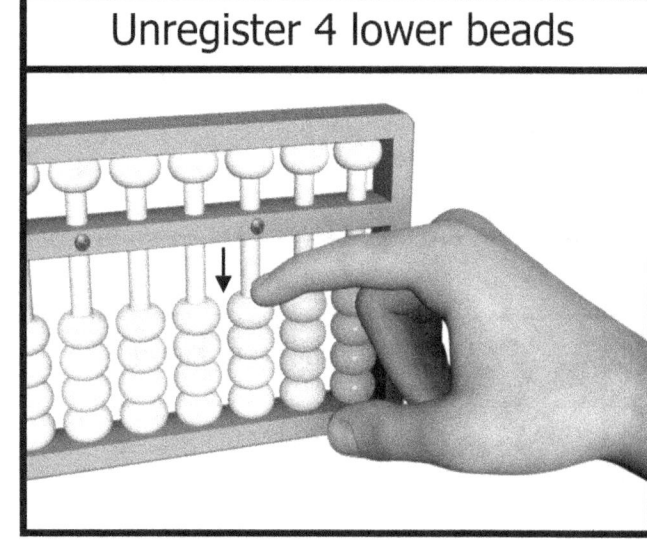
Unregister 1 upper bead	Register 1 upper bead
	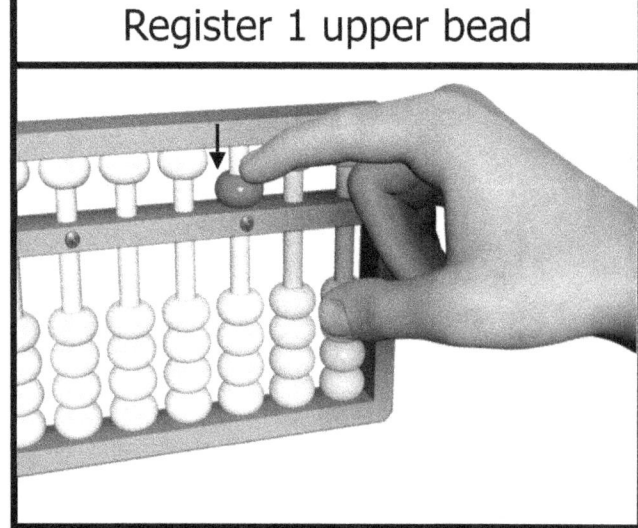

Time to use the **workbook!** Go to workbook **page 14**.

Addition

Addition is adding numbers to get the sum of those numbers

Addition - things to remember:
- Register your numbers from left to right, for example: for number 21 register the 2 first, and 1 last.
- Each digit must be registered in the correct column, for example with 21 the 2 for column 4 (tens column) and the 1 for column 3 (ones column).

Example: 21 + 12

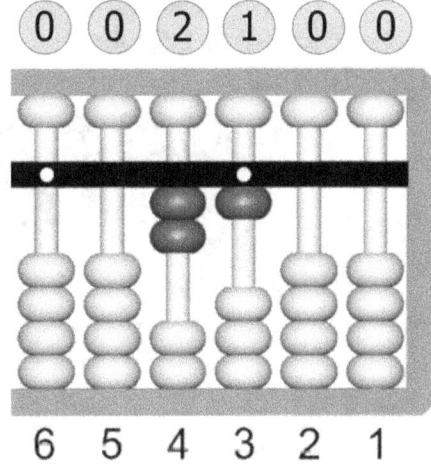

We will **register 21**

20
1
- Column 4, register 2 lower beads
- Column 3, register 1 lower bead

The abacus reads 21

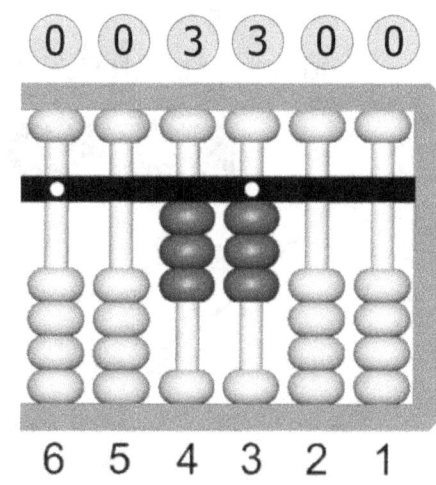

We will now **add 12** to 21

+10
+2
- Column 4, register 1 lower bead to add 10
- Column 3, register 2 lower beads to add 2

The abacus result is 33

These columns are useful to see the amount that you are adding. For example:
  20 means that you have just registered 20
  +2 means that you have just added 2

More addition examples

Example: 3 + 5

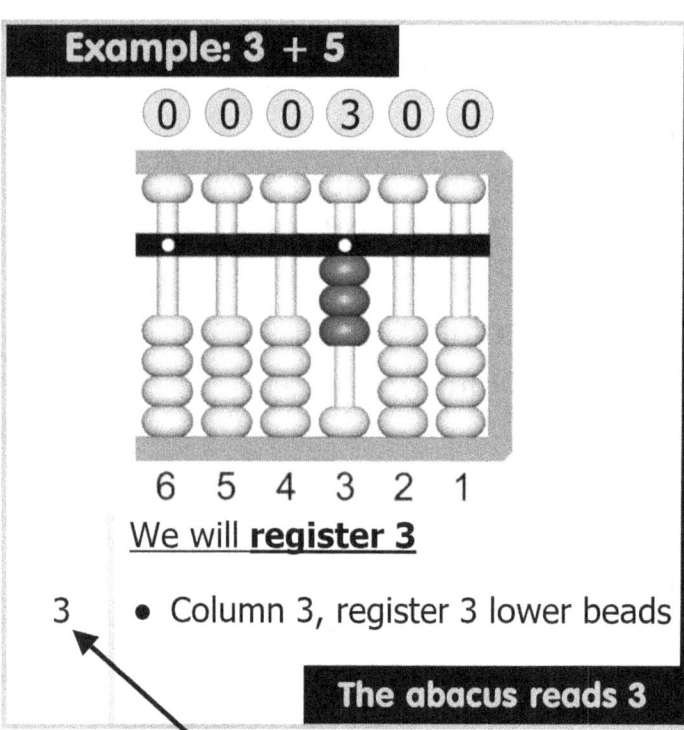

We will **register 3**

3   • Column 3, register 3 lower beads

The abacus reads 3

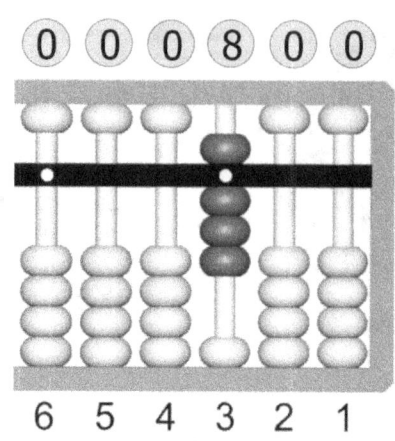

We will now **add 5** to 3

+5   • Column 3, register 1 upper bead to add 5

The abacus result is 8

> Remember!
> 3 means that you have just registered 3
> +5 means that you have just added 5

Example: 51 + 33

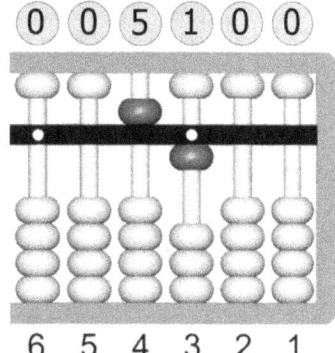

We will **register 51**

50   • Column 4, register 1 upper bead (this is the 5 of the 51)
1   • Column 3, register 1 lower bead (this is the 1 of the 51)

The abacus reads 51

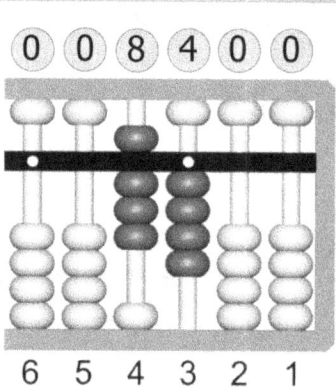

We will now **add 33** to 51

+30   • Column 4, register 3 lower beads to add 30
+3   • Column 3, register 3 lower beads to add 3

The abacus result is 84

More addition examples

Example: 36 + 52

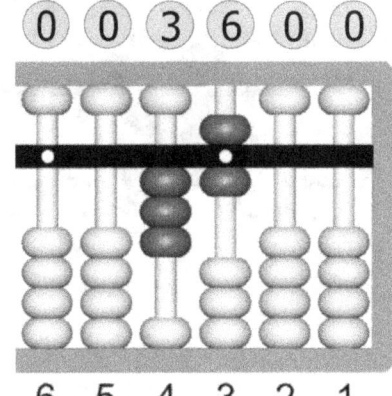

We will **register 36**

- 30 • Column 4, register 3 lower beads (this is the 3 of the 36)
- 6 • Column 3, register 1 upper bead and 1 lower bead (this is the 6 of the 36)

The abacus reads 36

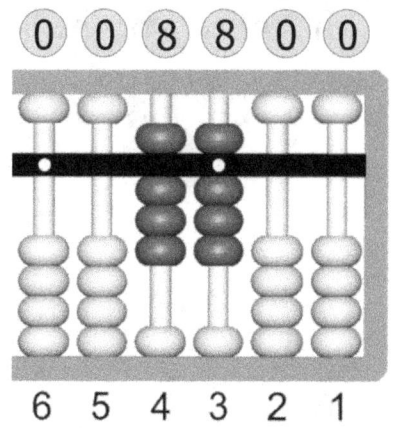

We will now **add 52** to 36

- +50 • Column 4, register 1 upper bead to add 50
- +2 • Column 3, register 2 lower beads to add 2

The abacus result is 88

Example: 11 + 36

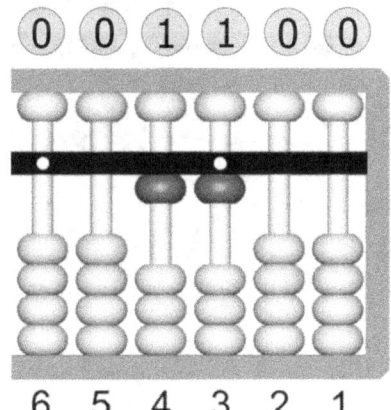

We will **register 11**

- 10 • Column 4, register 1 lower bead
- 1 • Column 3, register 1 lower bead

The abacus reads 11

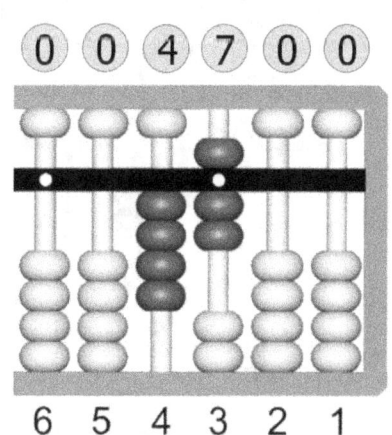

We will **add 36**

- +30 • Column 4, register 3 lower beads to add 30
- +6 • Column 3, register 1 upper bead and 1 lower bead to add 6

The abacus result is 47

More addition examples

Example: 60 + 29

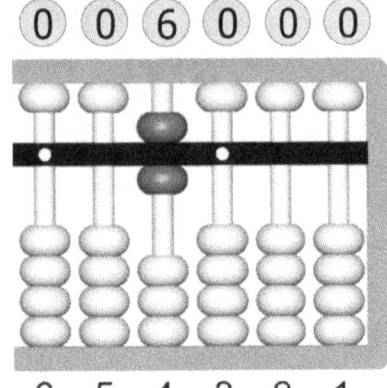

We will **register 60**

60
- Column 4, register 1 upper bead and 1 lower bead
  (this is the 6 of the 60)

- Column 3, do nothing

The abacus reads 60

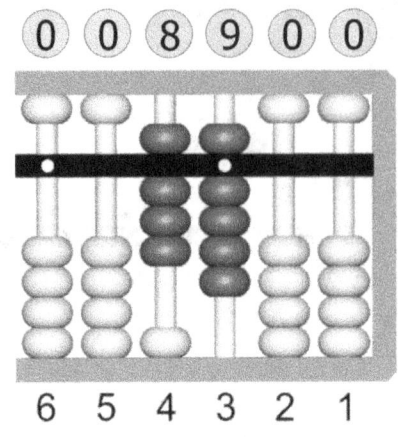

We will now **add 29** to 60

+20
- Column 4, register 2 lower beads to add 20

+9
- Column 3, register 1 upper bead and 4 lower beads to add 9
  (this is the 9 of the 29)

The abacus result is 89

Example: 79 + 10

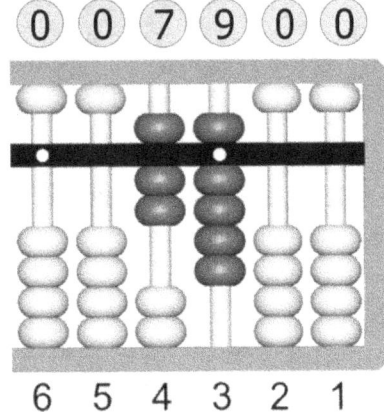

We will **register 79**

70
- Column 4, register 1 upper bead and 2 lower beads
  (this is the 7 of the 79)

9
- Column 3, register 1 upper bead and 4 lower beads
  (this is the 9 of the 79)

The abacus reads 79

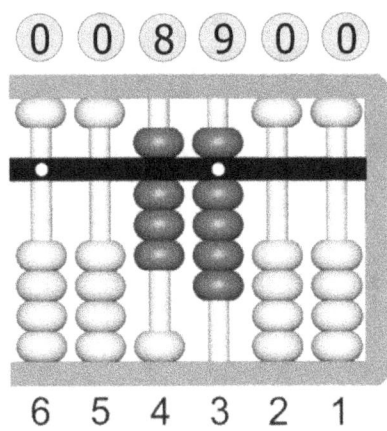

We will now **add 10** to 79

+10
- Column 4, register 1 lower bead to add 10
- Column 3, do nothing

The abacus result is 89

Addition when registering and unregistering in the same column

Sometimes we need to unregister and register in the same column.
For example if we need to add 3 beads to an already registered 4 lower beads to make 7, we need to register 1 upper and unregister 2 lower beads (+5-2=3).
Here are some examples.

Example: 54 + 13

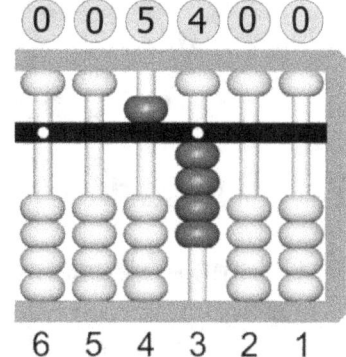

We will **register 54**

- 50 • Column 4, register 1 upper bead
- 4 • Column 3, register 4 lower beads

The abacus reads 54

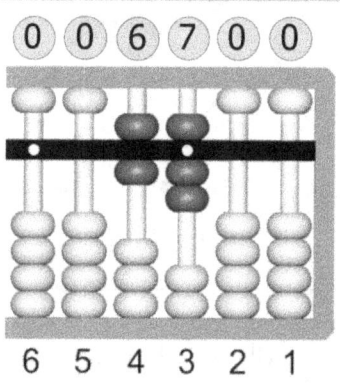

We will now **add 13** to 54

- +10 • Column 4, register 1 lower bead
- +3 • Column 3, register 1 upper bead (+5) and unregister 2 lower beads (-2)

The abacus result is 67

Example: 74 + 12

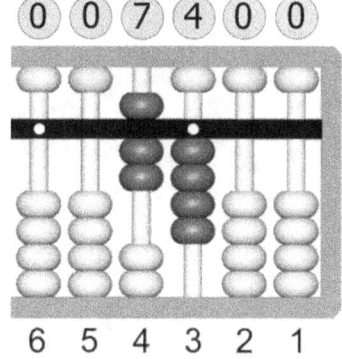

We will **register 74**

- 70 • Column 4, register 1 upper bead and 2 lower beads
- 4 • Column 3, register 4 lower beads

The abacus reads 74

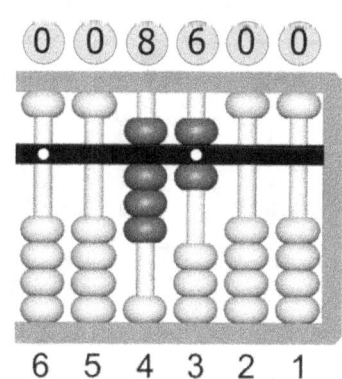

We will now **add 12** to 74

- +10 • Column 4, register 1 lower bead
- +2 • Column 3, register 1 upper bead (+5) and unregister 3 lower beads (-3)
(Total added in column 3 is 5-3=2)

The abacus result is 86

More addition examples

Example: 23 + 14

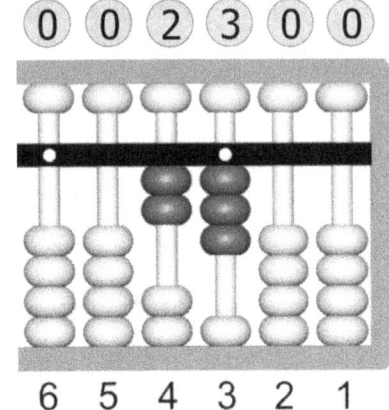

We will **register 23**

- 20 • Column 4, register 2 lower beads
  (this is the 2 of the 23)
- 3 • Column 3, register 3 lower beads
  (this is the 3 of the 23)

The abacus reads 23

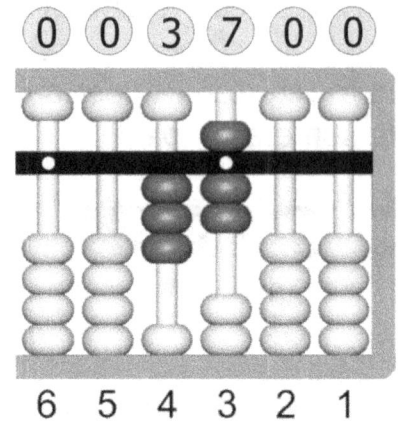

We will now **add 14** to 23

- +10 • Column 4, register 1 lower bead to add 10
- +4 • Column 3, register 1 upper bead (+5) and unregister 1 lower bead (-1)

The abacus result is 37

Example: 64 + 32

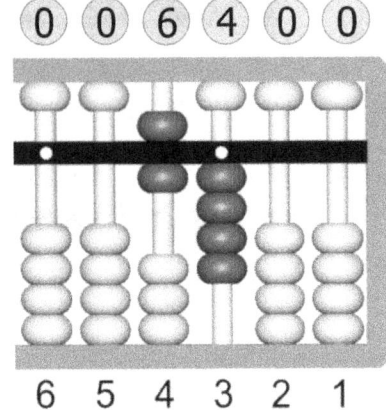

We will **register 64**

- 60 • Column 4, register 1 upper bead and 1 lower bead
  (this is the 6 of the 64)
- 4 • Column 3, register 4 lower beads
  (this is the 4 of the 64)

The abacus reads 64

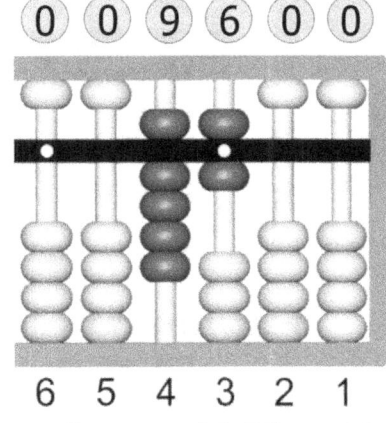

We will now **add 32** to 64

- +30 • Column 4, register 3 lower beads to add 30
- +2 • Column 3, register 1 upper bead (+5) and unregister 3 lower beads (-3)

The abacus result is 96

Time to use the **workbook**! Go to workbook **page 18**.

Imaginary abacus

Part 4

Doing mental math using an imaginary abacus rather than using an actual abacus can be achieved with practise. It is possible to learn to add and subtract without the need for the physical device.

When you imagine an abacus, it is best to imagine it with only the beam and the beads.

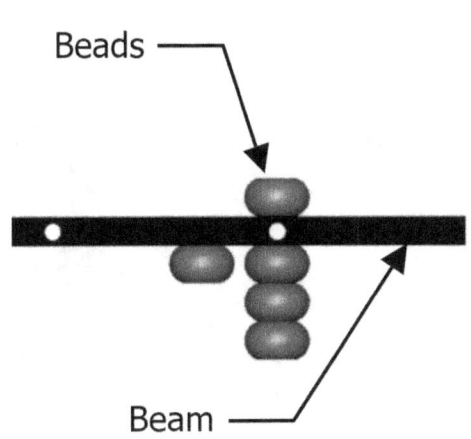

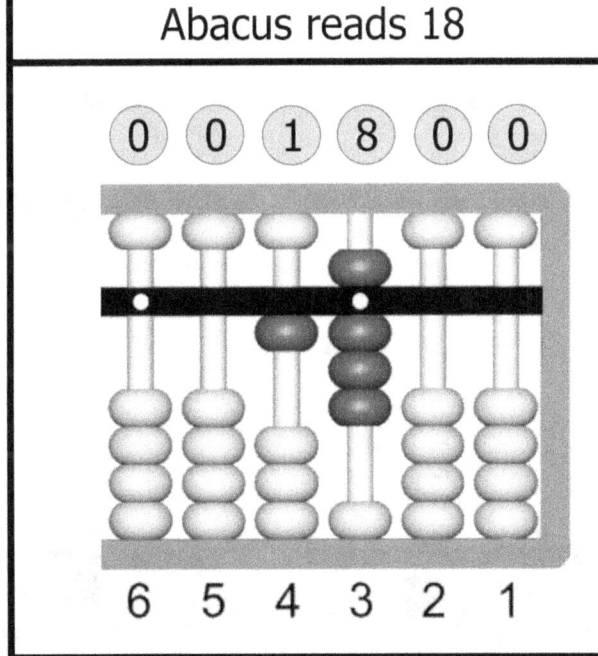

Abacus reads 18

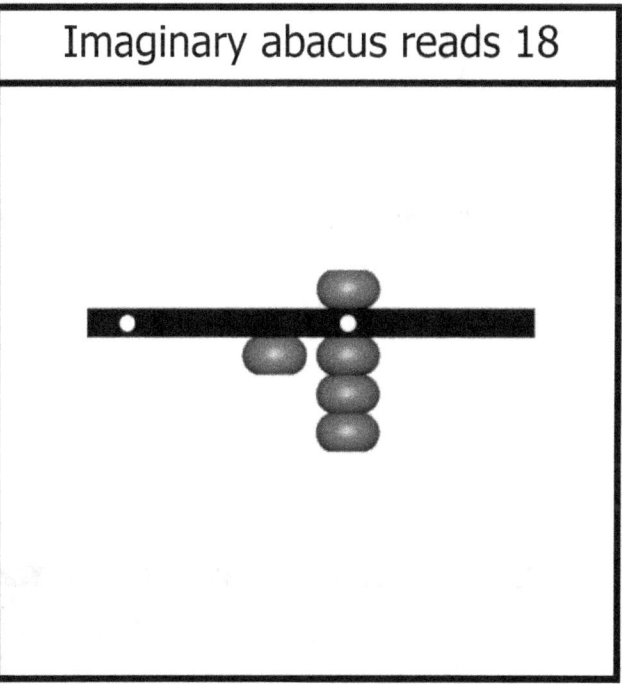

Imaginary abacus reads 18

Important!

- **Imagine** only the beads that are touching the beam, ignore all other beads!

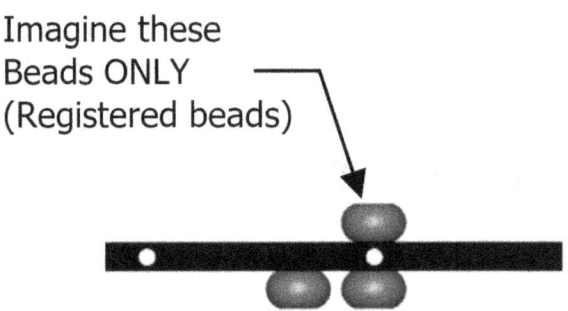

Imagine these Beads ONLY (Registered beads)

- Move your **fingers** in the air, just like you are using the physical abacus.

Some examples of what to imagine step-by-step.

Imagine

① Add 1 + 2

0	1	+2	=3

② Add 2 + 2

0	2	+2	=4

③ Add 5 + 2

0	5	+2	=7

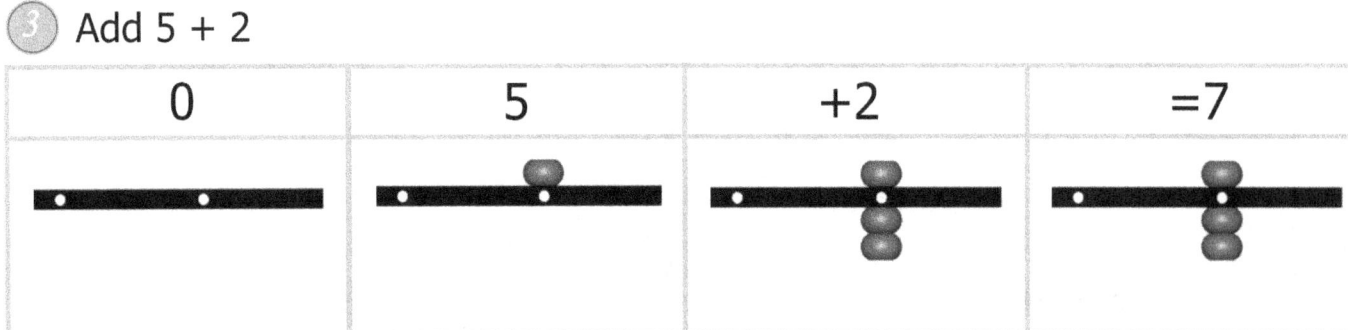

More examples of what to imagine step-by-step

④ Add 10 + 9

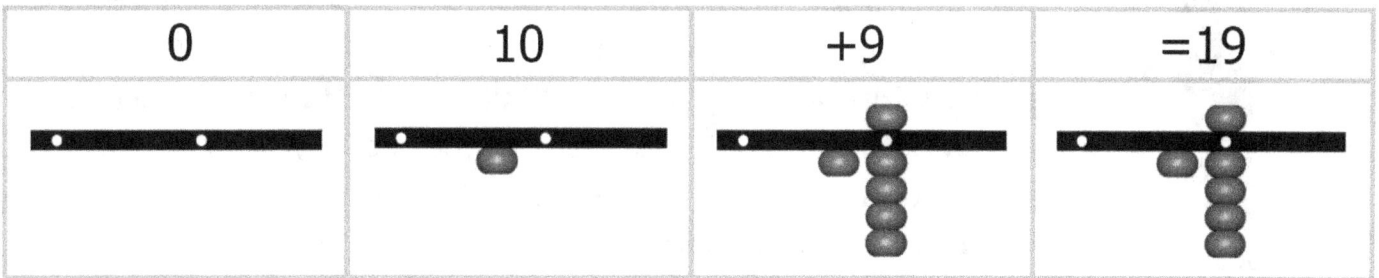

⑤ Add 35 + 13

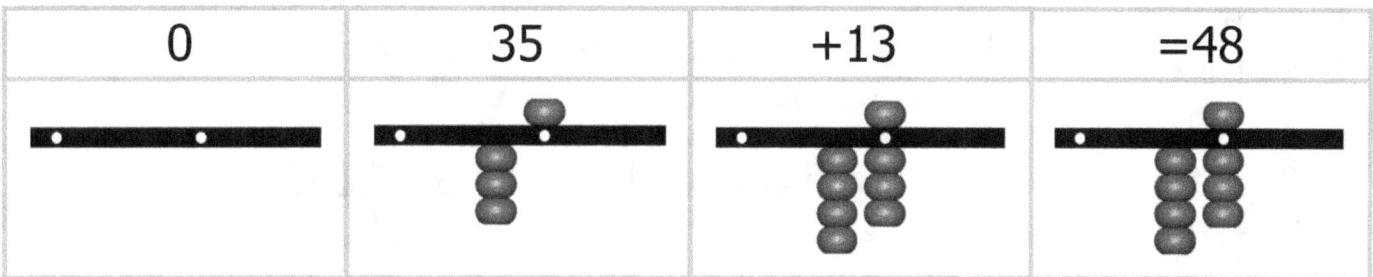

⑥ Add 62 + 22

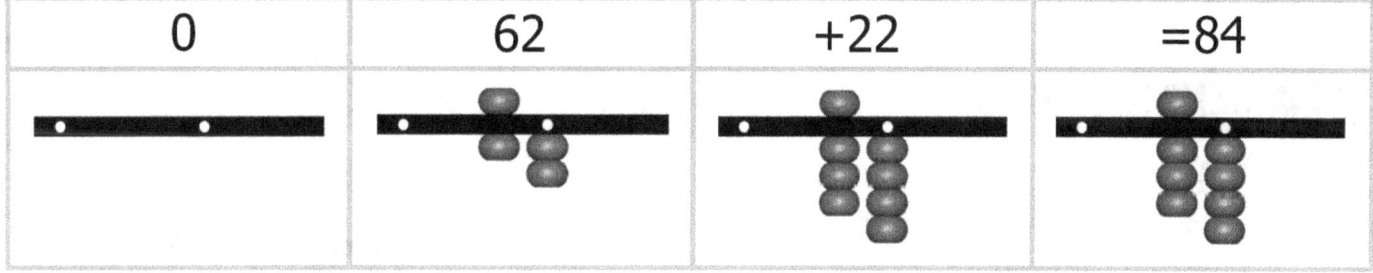

⑦ Add 70 + 20

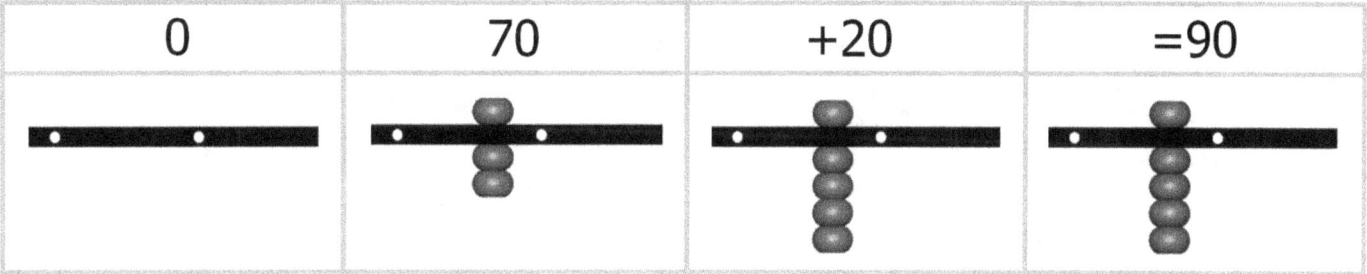

Not enough beads in the column for the addition

When you don't have enough beads, move to the next **LEFT** column to help

For example, when you try to add 4 to the already registered number 8, you don't have enough beads in the column to do it. You can only register a maximum of 9 in each column (4 lower beads and 1 upper bead, 4+5=9).
When this happens, we need to use the
'**Not enough beads list**'.

1=10-9
2=10-8
3=10-7
4=10-6
5=10-5
6=10-4
7=10-3
8=10-2
9=10-1

How to use the 'Not enough beads list'

Let's say we need to add 3 to a column but we don't have enough beads.

Look at the list, **3=10-7**

10 is the number to **register,** in the next **LEFT** column (1 lower bead).

7 is the number to **unregister** in our column.

Example: 8 + 3

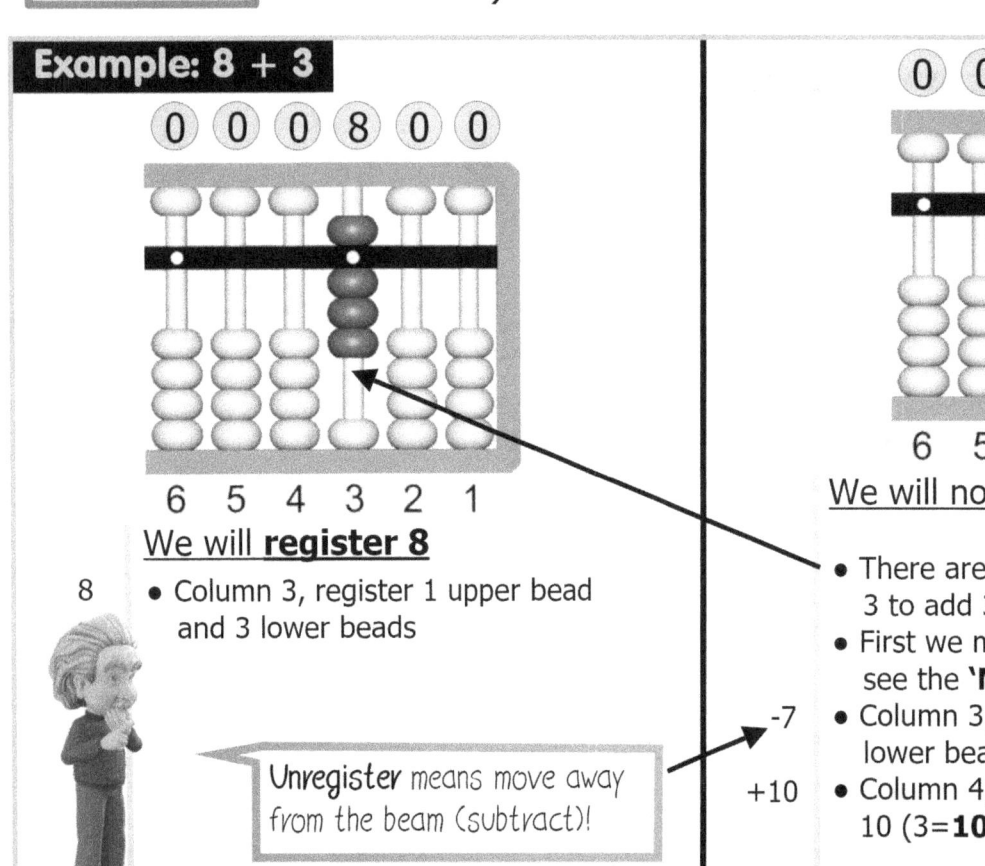

We will **register 8**

8 • Column 3, register 1 upper bead and 3 lower beads

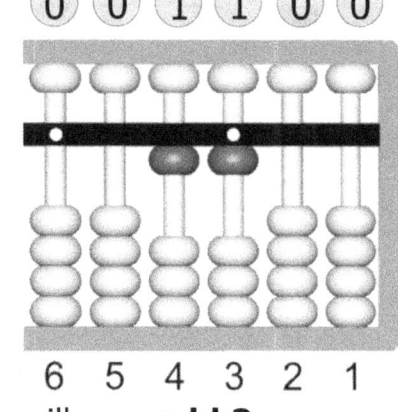

Unregister means move away from the beam (subtract)!

The abacus reads 8

We will now **add 3**

- There are not enough beads in column 3 to add 3
- First we must think that 3=**10-7** see the '**Not enough beads list**'
-7 • Column 3, **unregister** 1 upper and 2 lower beads to subtract 7 (3=10-**7**)
+10 • Column 4, register 1 lower bead to add 10 (3=**10**-7)

The abacus result is 11

More addition examples (when we don't have enough beads)

Example: 9 + 4

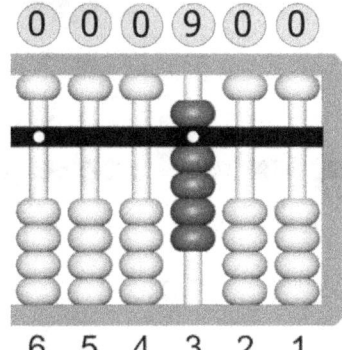

We will register 9

9
- Column 3, register 1 upper bead and 4 lower beads

The abacus reads 9

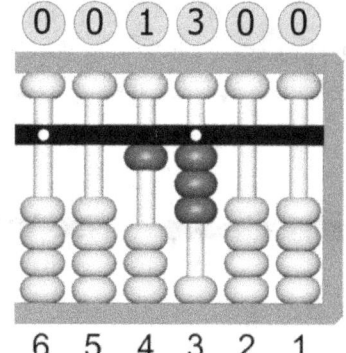

We will now add 4

There are not enough beads in column 3 to add 4, so think **4=10-6**, so remove 6 from column 3 then add 10 to column 4

-6
- Column 3, unregister 1 upper and 1 lower bead to subtract 6

+10
- Column 4, register 1 lower bead to add 10

The abacus result is 13

Example: 29 + 52

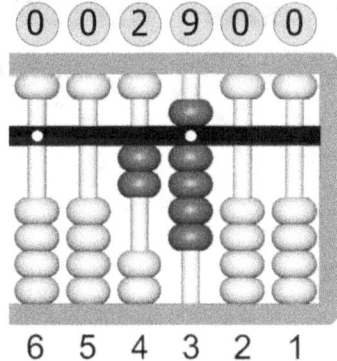

We will register 29

20
- Column 4, register 2 lower beads

9
- Column 3, register 1 upper bead and 4 lower beads

The abacus reads 29

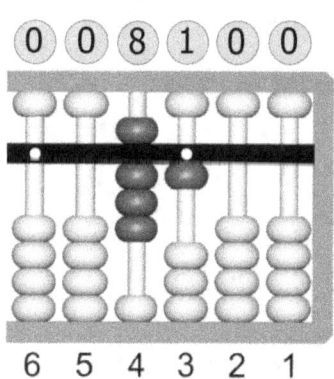

We will add 52

+50
- Column 4, register 1 upper bead (+50)

There are not enough beads in column 3 to register 2 more (to add 2), so think **2=10-8**

+10
- Column 4, register 1 lower bead

-8
- Column 3, unregister 1 upper and 3 lower beads

(Total from columns 4 & 3 is 10-8=2)

The abacus result is 81

38

Some examples of what to imagine step-by-step

① Add 8 + 3

0	8	+3	=11

② Add 18 + 4

0	18	+4	=22

③ Add 48 + 9

0	48	+9	=57

④ Add 25 + 15

0	25	+15	=40

⑤ Add 68 + 9

0	68	+9	=77

Some examples of what to imagine step-by-step

39 Imagine

⑥ Add 69 + 19

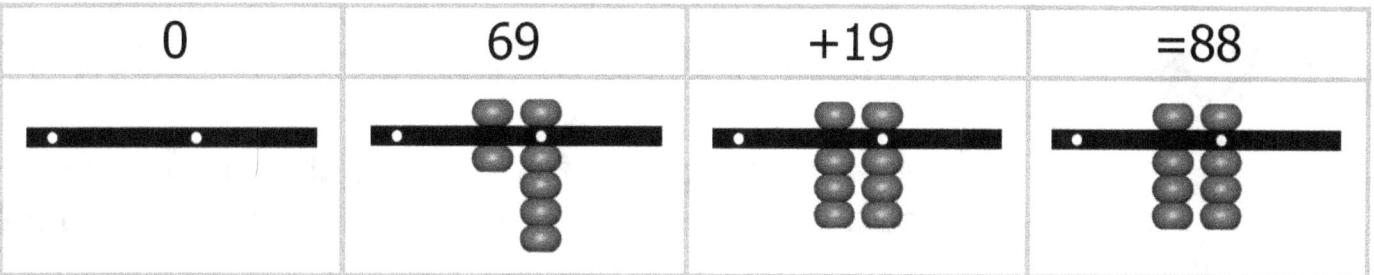

⑦ Add 17 + 16

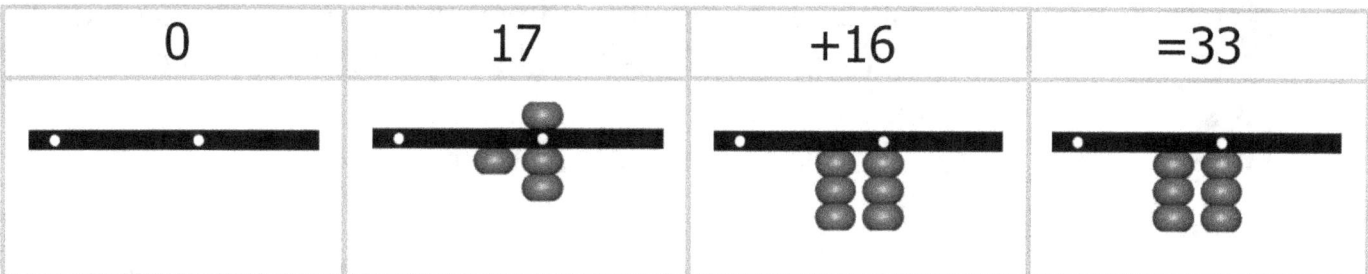

⑧ Add 24 + 9

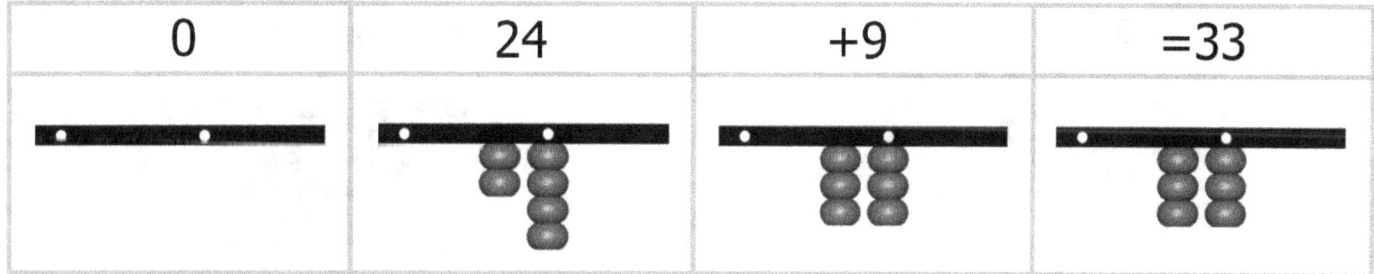

⑨ Add 75 + 15

| 0 | 75 | +15 | =90 |

Time to use the **workbook**! Go to workbook **page 23**.

Addition of 3 or more digit numbers

Part 5

Addition of larger numbers is no harder than adding small numbers. We just need to start adding in the leftmost column first.

Addition - things to remember:
- Register your numbers from left to right, for example: for number 971 register the 9 first, the 7 second and the 1 last.
- Each digit must be registered in the correct column, for example with 971, the 9 in column 5 (hundreds column), the 7 in column 4 (tens column) and the 1 in column 3 (ones column).
- Add the digits from left to right.

Example: 326 + 163

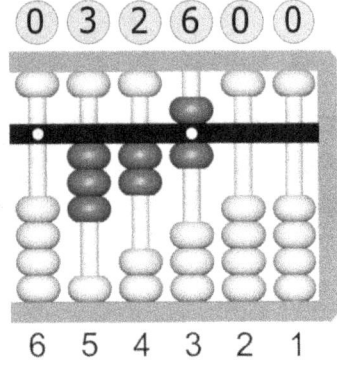

6 5 4 3 2 1

We will register 326

- 300 — Column 5, register 3 lower beads (this is the 3 of the 326)
- 20 — Column 4, register 2 lower beads (this is the 2 of the 326)
- 6 — Column 3, register 1 upper bead and 1 lower bead (this is the 6 of the 326)

The abacus reads 326

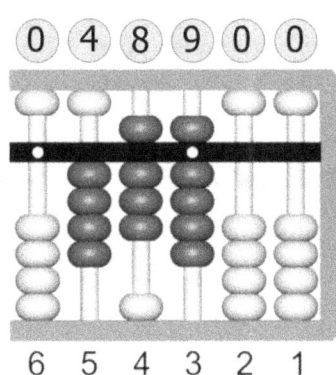

6 5 4 3 2 1

We will now add 163 to 326

- +100 — Column 5, register 1 lower bead to add 100
- +60 — Column 4, register 1 upper and 1 lower bead to add 60
- +3 — Column 3, register 3 lower beads to add 3

The abacus result is 489

Example: 2331 + 2103

(2)(3)(3)(1)(0)(0)

6 5 4 3 2 1

We will register 2331

2000	• Column 6, register 2 lower beads
300	• Column 5, register 3 lower beads
30	• Column 4, register 3 lower beads
1	• Column 3, register 1 lower bead

The abacus reads 2331

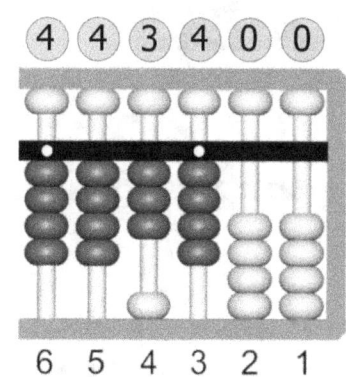

(4)(4)(3)(4)(0)(0)

6 5 4 3 2 1

We will add 2103

+2000	• Column 6, register 2 lower beads
+100	• Column 5, register 1 lower bead
	• Column 4, do nothing
+3	• Column 3, register 3 lower beads

The abacus result is 4434

Example: 7436 + 932

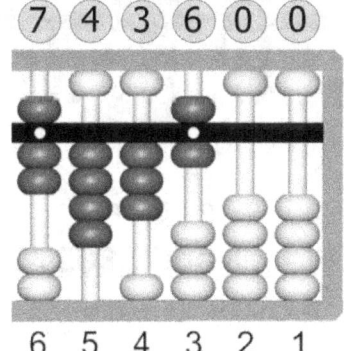

(7)(4)(3)(6)(0)(0)

6 5 4 3 2 1

We will register 7436

7000	• Column 6, register 1 upper bead and 2 lower beads
400	• Column 5, register 4 lower beads
30	• Column 4 register 3 lower beads
6	• Column 3, register 1 upper bead and 1 lower bead

The abacus reads 7436

(8)(3)(6)(8)(0)(0)

6 5 4 3 2 1

We will now add 932

There are not enough beads in column 5 to register 9 more, so think **9=10-1**

-100	• Column 5, unregister 1 lower bead
+1000	• Column 6, register 1 lower bead
+30	• Column 4, register 1 upper bead and unregister 2 lower beads
+2	• Column 3, register 2 lower beads

The abacus result is 8368

More addition examples (when we don't have enough beads)

Example: 45 + 5

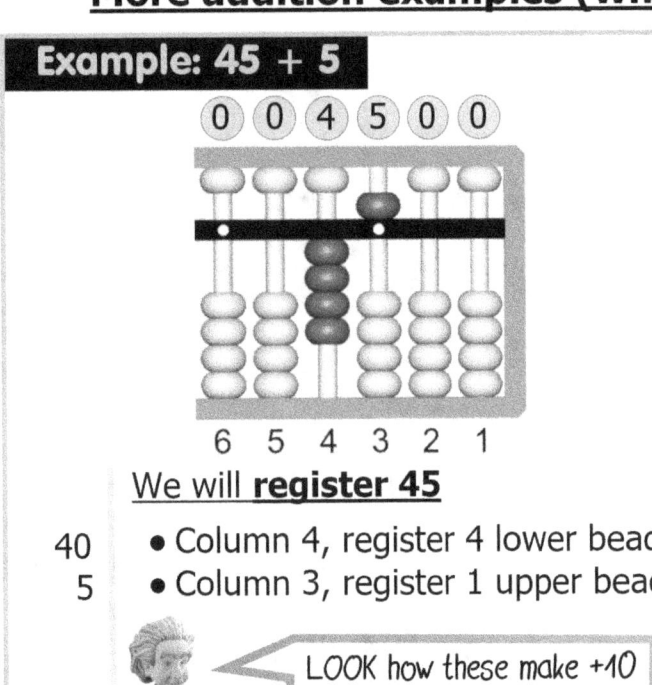

We will **register 45**

- 40 — Column 4, register 4 lower beads
- 5 — Column 3, register 1 upper bead

LOOK how these make +10
+50−40=+10

The abacus reads 45

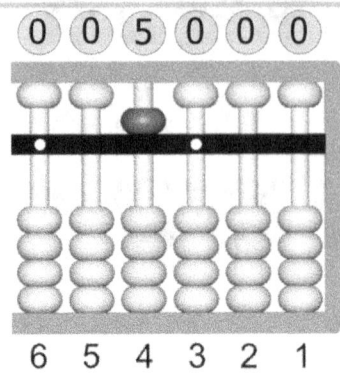

We will **add 5**

There are not enough beads in column 3 to register 5 more (to add 5), so think **5=10-5**

- −5 — Column 3, unregister 1 upper bead
- +50 — Column 4, register 1 upper bead
- −40 — Column 4, unregister 4 lower beads

The abacus result is 50

Example: 5395607 + 2803721

We will **register 5395607**

- 5000000 — Column 9, register 1 upper bead
- 300000 — Column 8, register 3 lower beads
- 90000 — Column 7, register 1 upper bead and 4 lower beads
- 5000 — Column 6, register 1 upper bead
- 600 — Column 5, register 1 upper bead and 1 lower bead
- Column 4, do nothing
- 7 — Column 3, register 1 upper bead and 2 lower beads

The abacus reads 5395607

We will **add 2803721**

- +2000000 — Column 9, register 2 lower beads

There are not enough beads in column 8 to register 8 more, so think **8=10-2**
- −200000 — Column 8, unregister 2 lower beads
- +1000000 — Column 9, register 1 lower bead

- Column 7, do nothing
- +3000 — Column 6, register 3 lower beads

There are not enough beads in column 5 to register 7 more, so think **7=10-3**
- −500 — Column 5, unregister 1 upper bead
- +200 — Column 5, register 2 lower beads
- +1000 — Column 6, register 1 lower bead

- +20 — Column 4, register 2 lower beads
- +1 — Column 3, register 1 lower bead

The abacus result is 8199328

More addition examples (when we don't have enough beads)

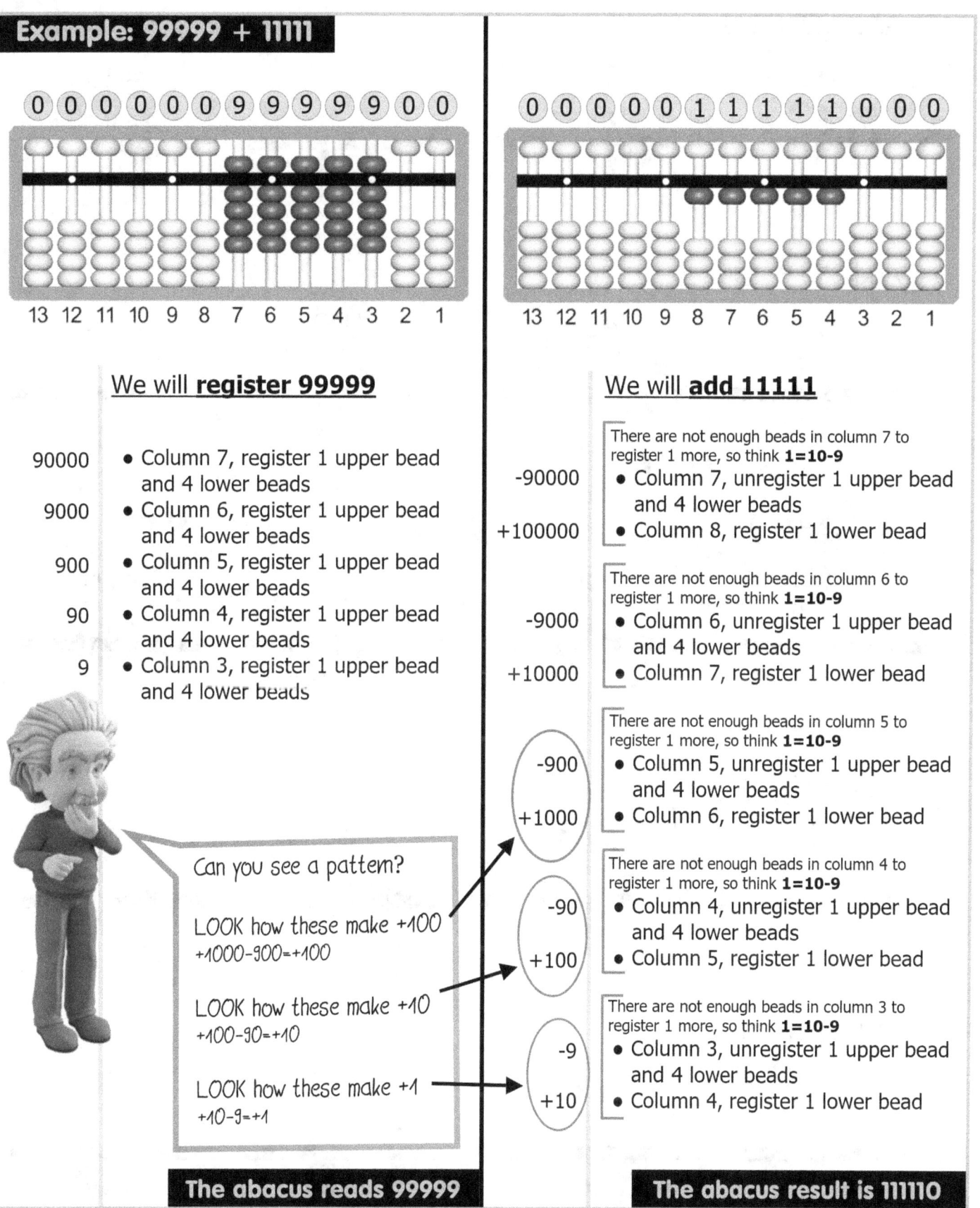

44

Some examples of what to imagine step-by-step

① Add 122 + 111

| 0 | 122 | +111 | =233 |

② Add 453 + 100

| 0 | 453 | +100 | =553 |

③ Add 754 + 13

| 0 | 754 | +13 | =767 |

④ Add 903 + 53

| 0 | 903 | +53 | =956 |

⑤ Add 426 + 123

| 0 | 426 | +123 | =549 |

Some examples of what to imagine step-by-step

6) Add 224 + 112

0	224	+112	=336

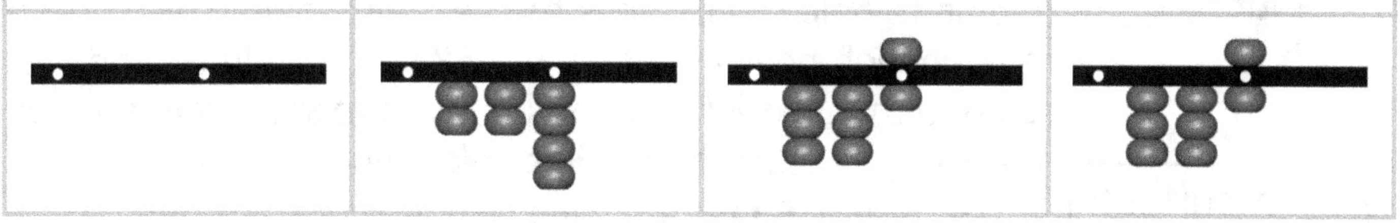

7) Add 846 + 123

0	846	+123	=969

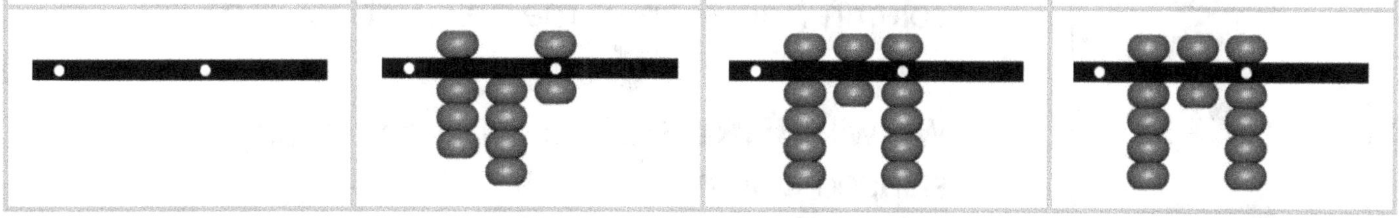

8) Add 662 + 125

0	662	+125	=787

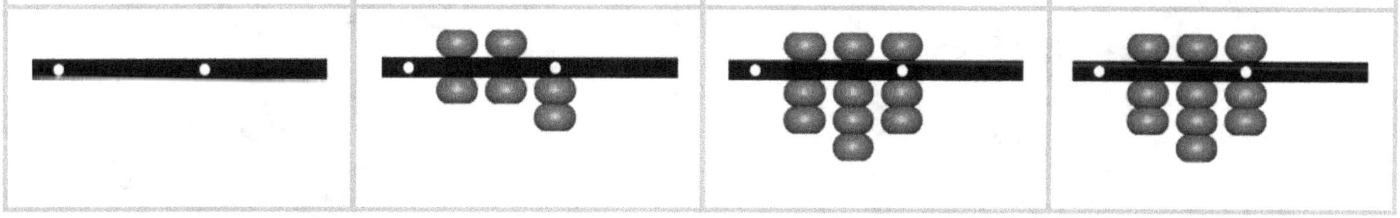

9) Add 521 + 155

0	521	+155	=676

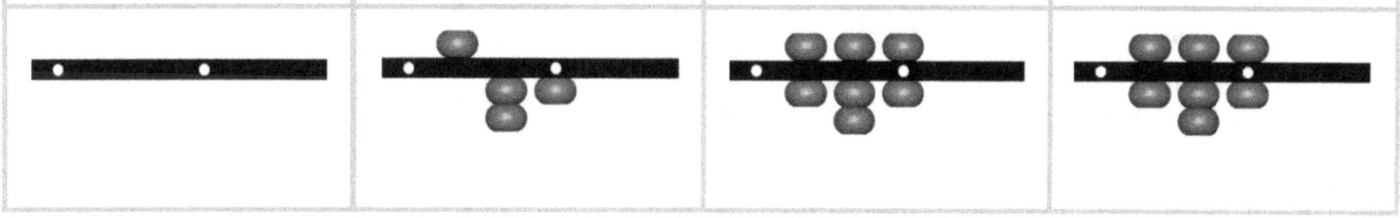

Time to use the **workbook**! Go to workbook **page 30**.

Skipped columns when adding

Part 6

Sometimes we have to SKIP a column. I'll explain why below.

We've learnt that when a column doesn't have enough beads left on it to make the addition, we move to the next LEFT column to help. Sometimes the next left column also doesn't have enough beads on it, so we **SKIP** this column and move again to the next left column until you reach a column that has enough beads to use. See below how it works.

Important!

- We will **SKIP** a column when there are not enough beads to use in that column.
- We will see this symbol when we need to skip a column (move on to the next left column).

⬅

- We will **UNREGISTER** all beads in any skipped columns.

Example: 295 + 9

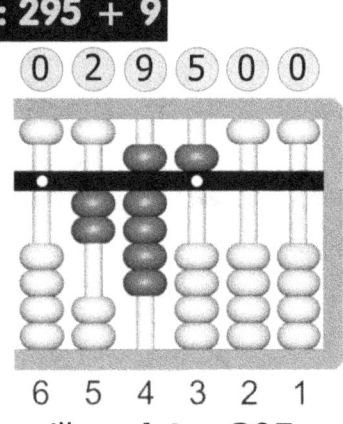

We will **register 295**

- 200 • Column 5, register 2 lower beads
- 90 • Column 4, register 1 upper bead and 4 lower beads
- 5 • Column 3, register 1 upper bead

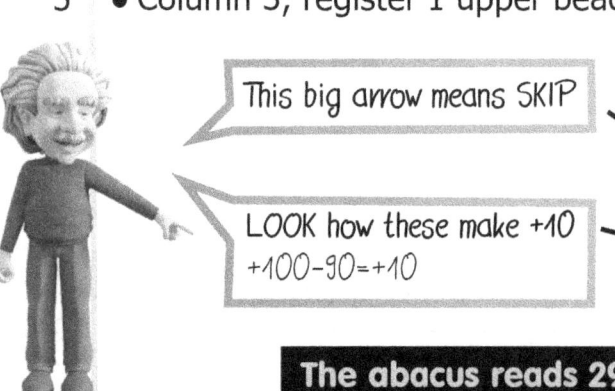

This big arrow means SKIP

LOOK how these make +10
+100-90=+10

The abacus reads 295

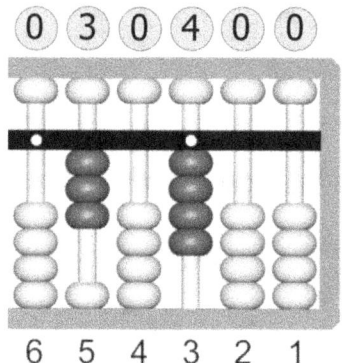

We will now **add 9**

There are not enough beads in column 3 to register 9 more, so think **9=10-1**

- −1 • Column 3, unregister 1 upper bead and register 4 lower beads

We would normally move to the next LEFT column and add 1 lower bead (to add 10) but we don't have 1 bead left to use

- −90 • Column 4, **SKIP** this column and unregister all beads
- +100 • Column 5, register 1 lower bead

The abacus result is 304

More skipped column examples

Example: 995 + 9

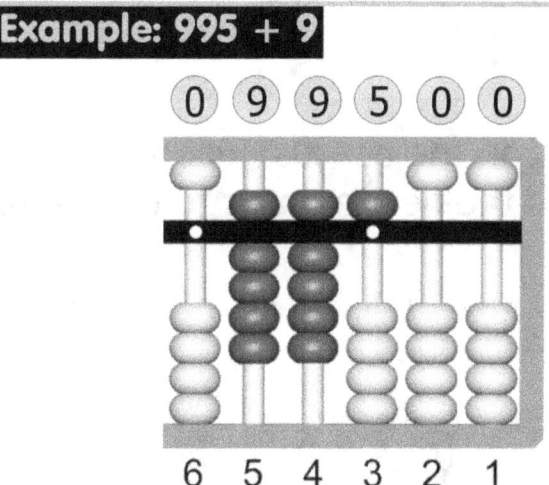

We will **register 995**

- 900 • Column 5, register 1 upper bead and 4 lower beads
- 90 • Column 4, register 1 upper bead and 4 lower beads
- 5 • Column 3, register 1 upper bead

The abacus reads 995

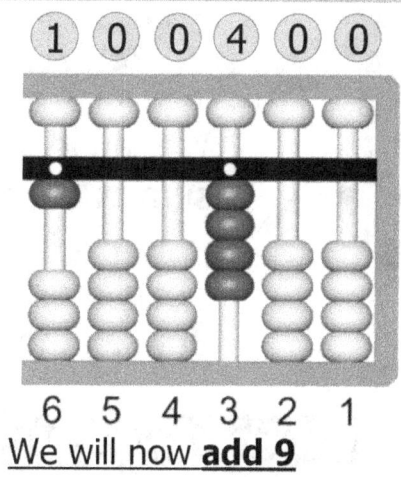

We will now **add 9**

There are not enough beads in column 3 to register 9 more, so think **9=10-1**

- −1 • Column 3, unregister 1 upper bead and register 4 lower beads
- ← • Column 4, **SKIP** this column and
- −90   unregister all beads
- ← • Column 5, **SKIP** this column and
- −900   unregister all beads
- +1000 • Column 6, register 1 lower bead

The abacus result is 1004

Example: 9999 + 1

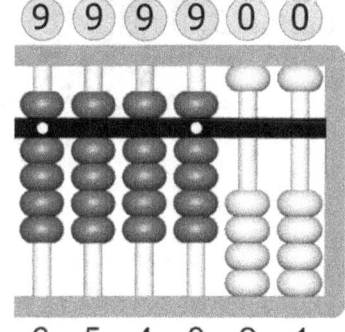

We will **register 9999**

- 9000 • Column 6, register 1 upper bead and 4 lower beads
- 900 • Column 5, register 1 upper bead and 4 lower beads
- 90 • Column 4, register 1 upper bead and 4 lower beads
- 9 • Column 3, register 1 upper bead and 4 lower beads

The abacus reads 9999

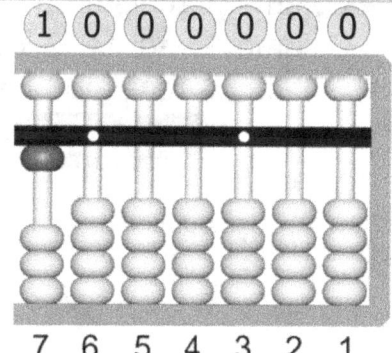

We will now **add 1**

There are not enough beads in column 3 to register 1 more, so think **1=10-9**

- −9 • Column 3, unregister all beads
- ← • Column 4, **SKIP** this column and
- −90   unregister all beads, move to column 5
- ← • Column 5, **SKIP** this column and
- −900   unregister all beads, move to column 6
- ← • Column 6, **SKIP** this column and
- −9000   unregister all beads, move to column 7
- +10000 • Column 7, register 1 lower bead

The abacus result is 10000

Addition of 3 or more numbers

Sometimes we have to add 3 or more numbers, here's how.

When we add many numbers on the abacus, just find the sum of the first two, then add the next number to that sum.

Keep adding one number to the sum of the previous numbers until all the numbers have been added.

Example: 223 + 235 + 511

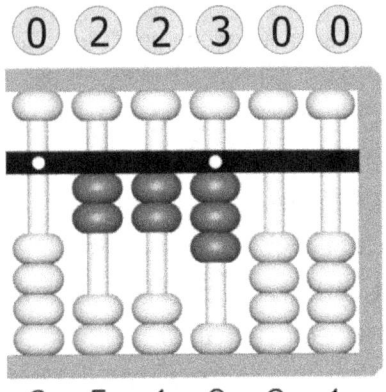

We will **register 223**

- 200 • Column 5, register 2 lower beads
- 20 • Column 4, register 2 lower beads
- 3 • Column 3, register 3 lower beads

The abacus reads 223

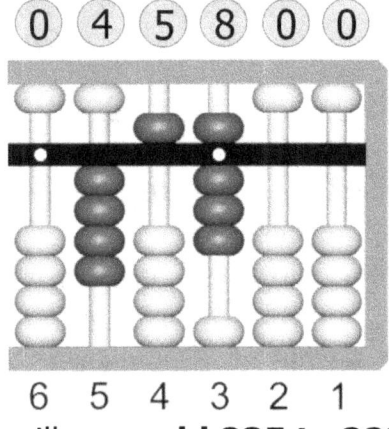

We will now **add 235 to 223**

- +200 • Column 5, register 2 lower beads
- +30 • Column 4, register 1 upper bead and unregister 2 lower beads
- +5 • Column 3, register 1 upper bead

The abacus sum is 458

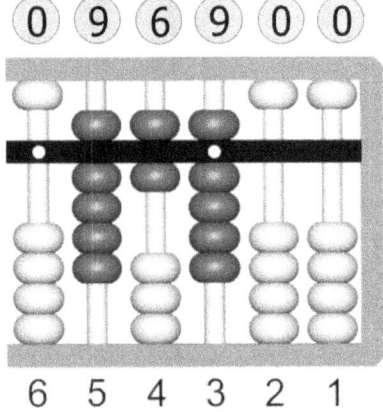

We will now **add 511 to the sum 458**

- +500 • Column 5, register 1 upper bead
- +10 • Column 4, register 1 lower bead
- +1 • Column 3, register 1 lower bead

The abacus result is 969

Addition of 3 or more numbers

Example: 615631 + 363160 + 1210

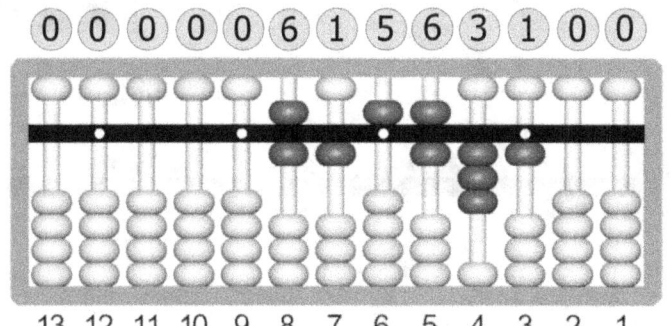

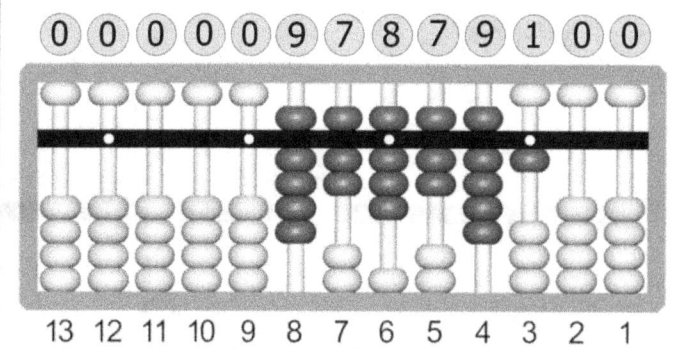

We will **register 615631**

600000	• Column 8, register 1 upper bead and 1 lower bead
10000	• Column 7, register 1 lower bead
5000	• Column 6, register 1 upper bead
600	• Column 5, register 1 upper bead and 1 lower bead
30	• Column 4, register 3 lower beads
1	• Column 3, register 1 lower bead

The abacus reads 615631

We will now **add 363160 to 615631**

+300000	• Column 8, register 3 lower beads
+60000	• Column 7, register 1 upper bead and 1 lower bead
+3000	• Column 6, register 3 lower beads
+100	• Column 5, register 1 lower bead
+60	• Column 4, register 1 upper bead and 1 lower bead
	• Column 3, do nothing

The abacus sum is 978791

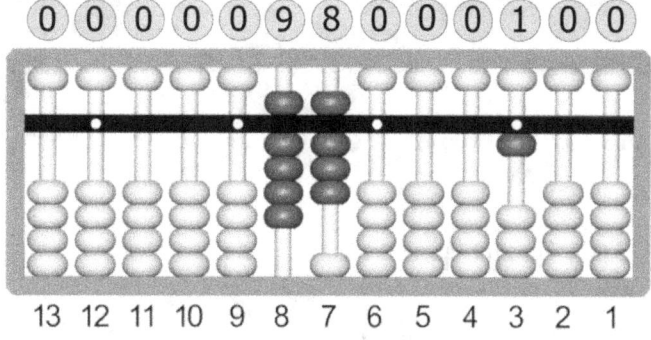

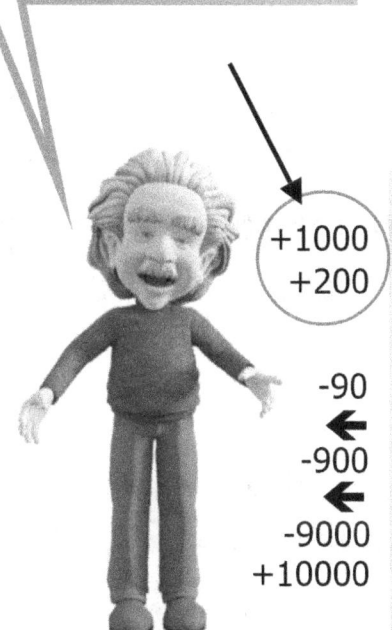

After we registered these, columns 6 & 5 had no beads left to use.

We will now **add 1210 to the sum 978791**

+1000	• Column 6, register 1 lower bead
+200	• Column 5, register 2 lower beads
	There are not enough beads in column 4 to register 1 more, so think **1=10-9**
−90	• Column 4, unregister all beads
← −900	• Column 5, **SKIP** this column and unregister all beads, move to column 6
← −9000	• Column 6, **SKIP** this column and unregister all beads, move to column 7
+10000	• Column 7, register 1 lower bead
	• Column 3, do nothing

The abacus result is 980001

Some examples of what to imagine step-by-step

① Add 96 + 9

| 0 | 96 | +9 | =105 |

② Add 99 + 2

| 0 | 99 | +2 | =101 |

③ Add 398 + 14

| 0 | 398 | +14 | =412 |

④ Add 403 + 40 + 4

| 0 | 403 | +40 | +4 | =447 |

⑤ Add 510 + 43 + 5

| 0 | 510 | +43 | +5 | =558 |

Time to use the **workbook**! Go to workbook **page 36**.

Subtraction

Part 7

Subtraction is taking one number away from another to find the difference.

Subtraction - things to remember:
- Register your numbers from left to right, just the same as we did with addition, for example:
for number 641 register the 6 first, 4 second and 1 last.
- Each digit must be registered in the correct column, for example with 641 the 6 is for column 5 (hundredths column), the 4 for column 4 (tens column) and the 1 for column 3 (ones column), just like we did with addition.

Example: 43 - 21

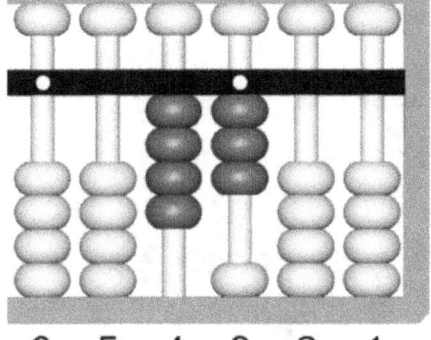

We will **register 43**

40 • Column 4, register 4 lower beads
3 • Column 3, register 3 lower beads

The abacus reads 43

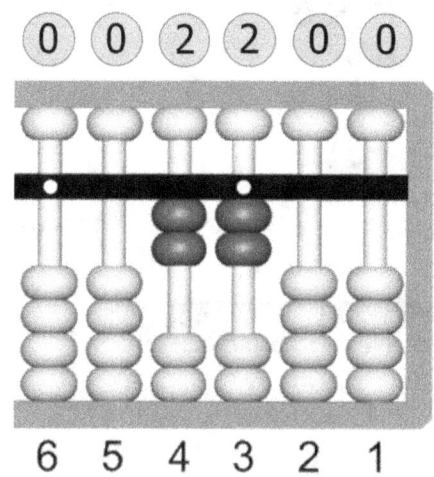

We will now **subtract 21** from 43

-20 • Column 4, unregister 2 lower beads to subtract 20
-1 • Column 3, unregister 1 lower bead to subtract 1

The abacus result is 22

These columns are useful to see the amount that you are subtracting.
For example:
 40 means that you have just registered 40
 -20 means that you have just subtracted 20

52 Subtracting numbers that have different amounts of digits

For example, when subtracting **543 - 21** we see that 543 has 3 digits and 21 only has 2.

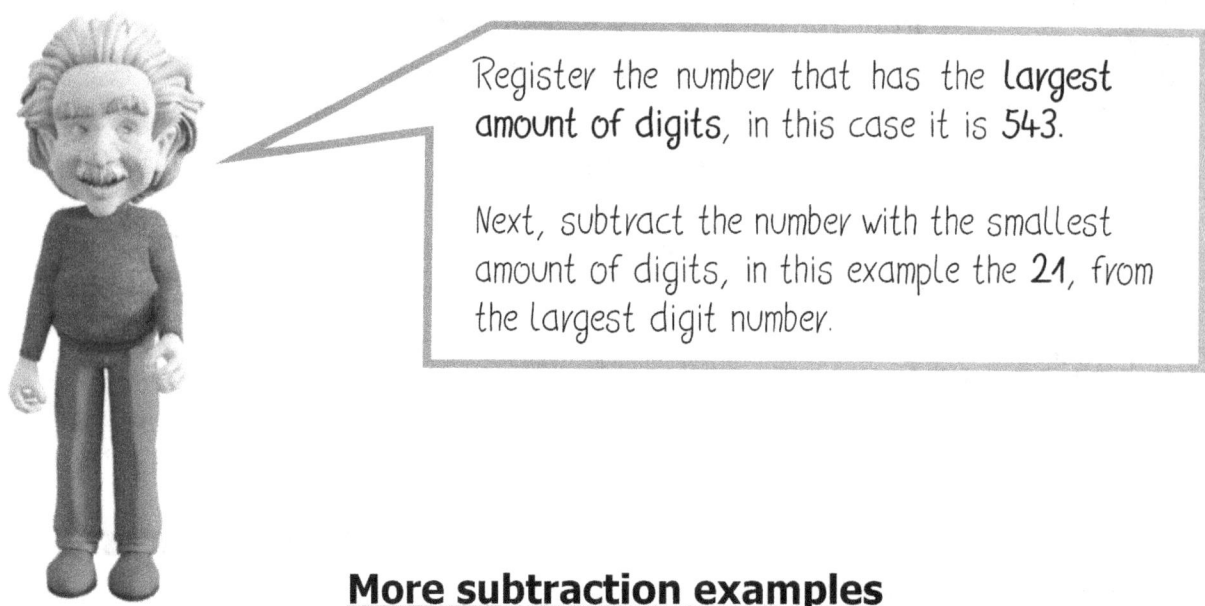

Register the number that has the **largest amount of digits**, in this case it is 543.

Next, subtract the number with the smallest amount of digits, in this example the 21, from the largest digit number.

More subtraction examples

Example: 543 - 21

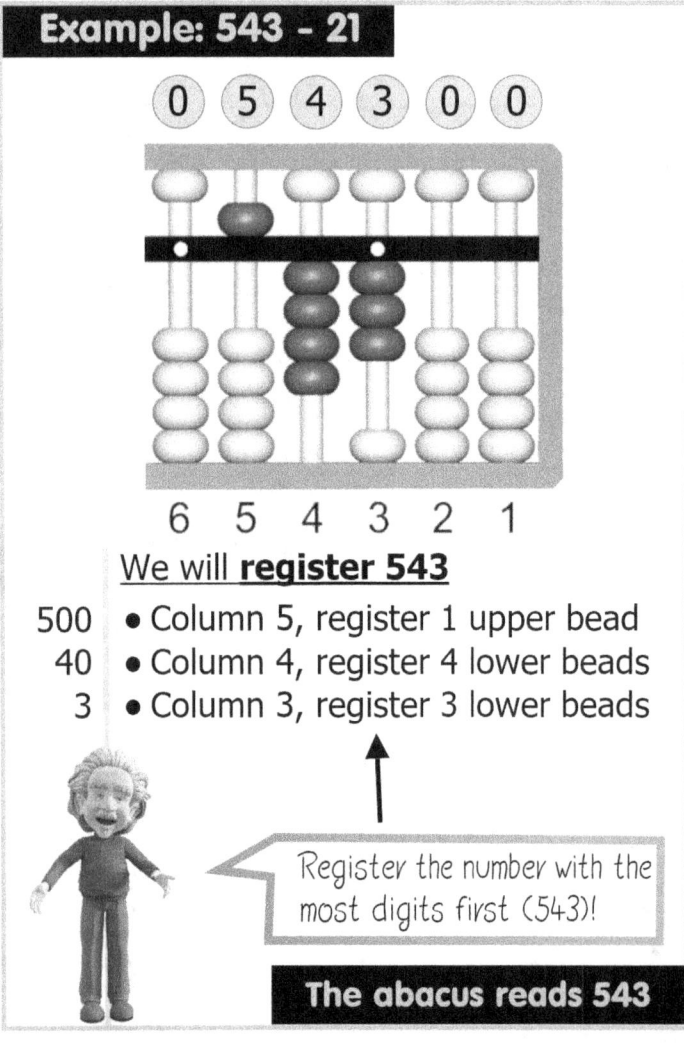

We will **register 543**

- 500 • Column 5, register 1 upper bead
- 40 • Column 4, register 4 lower beads
- 3 • Column 3, register 3 lower beads

Register the number with the most digits first (543)!

The abacus reads 543

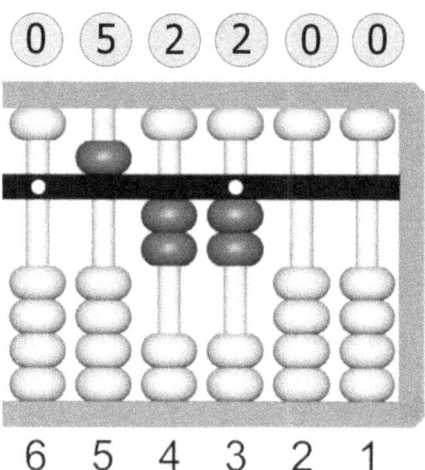

We will now **subtract 21**

- -20 • Column 4, unregister 2 lower beads to subtract 20
- -1 • Column 3, unregister 1 lower bead to subtract 1

The abacus result is 522

More subtraction examples

Example: 9 - 5

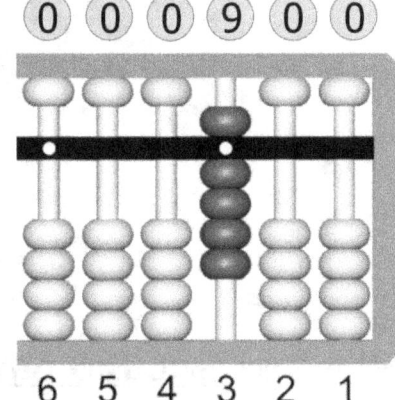

We will **register 9**

9 • Column 3, register 1 upper bead and 4 lower beads

The abacus reads 9

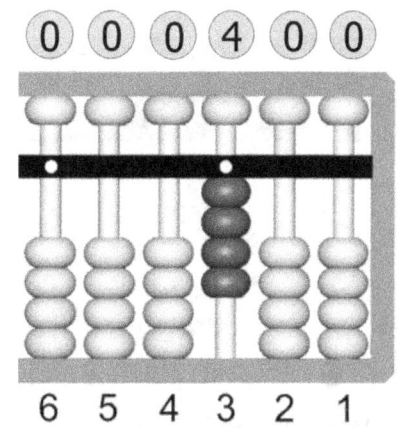

We will now **subtract 5**

-5 • Column 3, unregister 1 upper bead

The abacus result is 4

Example: 98 - 21

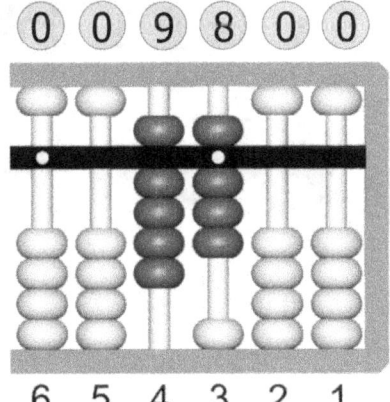

We will **register 98**

90 • Column 4, register 1 upper bead and 4 lower beads

8 • Column 3, register 1 upper bead and 3 lower beads

The abacus reads 98

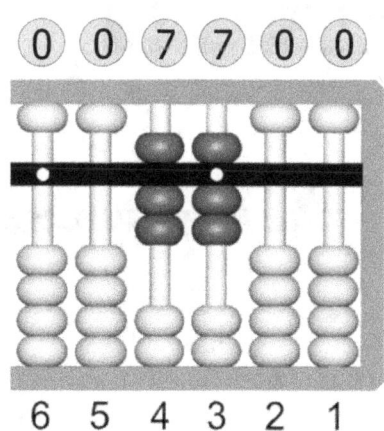

We will **subtract 21**

-20 • Column 4, unregister 2 lower beads
-1 • Column 3, unregister 1 lower bead

The abacus result is 77

More subtraction examples

Example: 686 - 661

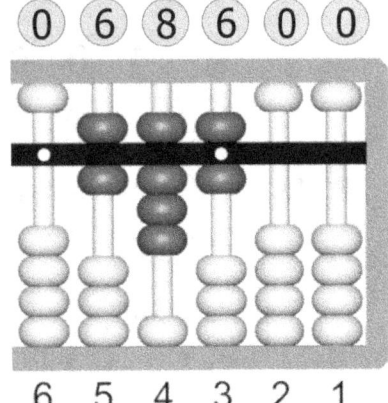

We will register 686

- 600 • Column 5, register 1 upper bead and 1 lower bead
- 80 • Column 4, register 1 upper bead and 3 lower beads
- 6 • Column 3, register 1 upper bead and 1 lower bead

The abacus reads 686

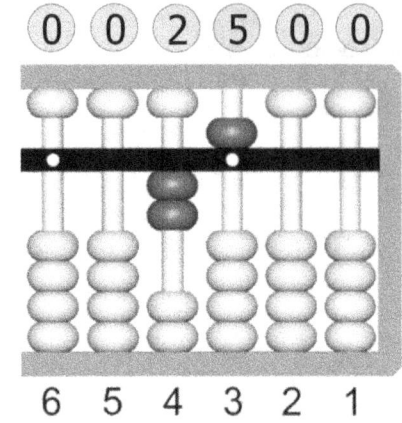

We will now subtract 661

- -600 • Column 5, unregister 1 upper bead and 1 lower bead
  (Total = -500-100=-600)
- -60 • Column 4, unregister 1 upper bead and 1 lower bead
  (Total = -50-10=-60)
- -1 • Column 3, unregister 1 lower bead

The abacus result is 25

Example: 8543 - 432

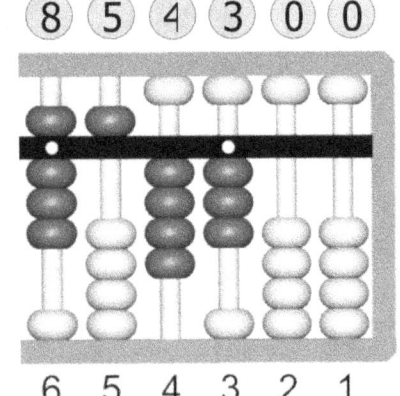

We will register 8543

- 8000 • Column 6, register 1 upper bead and 3 lower beads
- 500 • Column 5, register 1 upper bead
- 40 • Column 4, register 4 lower beads
- 3 • Column 3, register 3 lower beads

The abacus reads 8543

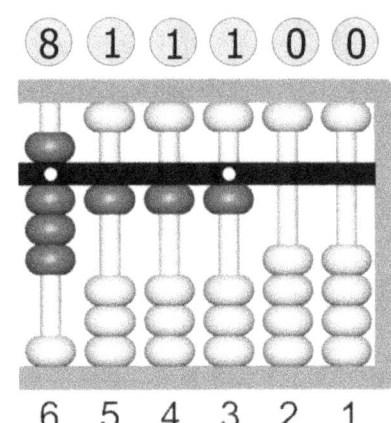

We will now subtract 432

- -400 • Column 5, unregister 1 upper bead and register 1 lower bead
  (Total = -500+100=-400)
- -30 • Column 4, unregister 3 lower beads
- -2 • Column 3, unregister 2 lower beads

The abacus result is 8111

More subtraction examples

Example: 205 - 102

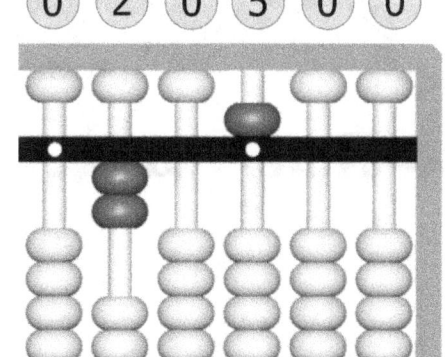

We will **register 205**

- 200 • Column 5, register 2 lower beads
- • Column 4, do nothing
- 5 • Column 3, register 1 upper bead

The abacus reads 205

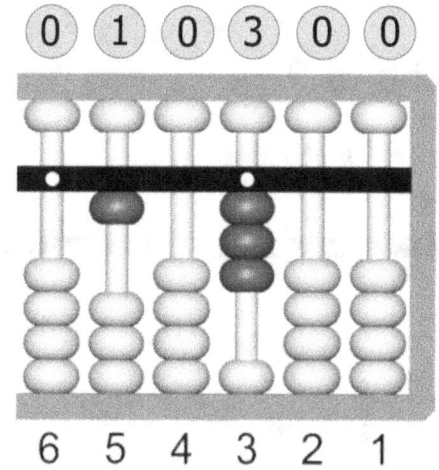

We will now **subtract 102**

- -100 • Column 5, unregister 1 lower bead
- • Column 4, do nothing
- -2 • Column 3, unregister 1 upper bead and register 3 lower beads

The abacus result is 103

Example 4374 - 2064

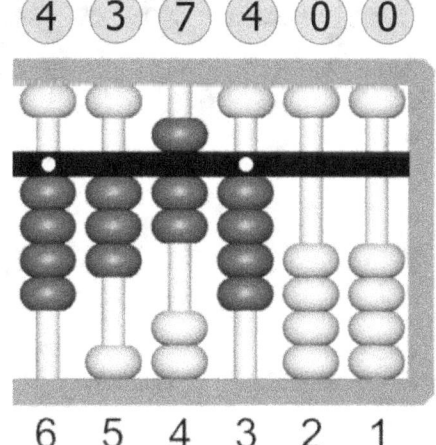

We will **register 4374**

- 4000 • Column 6, register 4 lower beads
- 300 • Column 5, register 3 lower beads
- 70 • Column 4, register 1 upper and 2 lower beads
- 4 • Column 3, register 4 lower beads

The abacus reads 4374

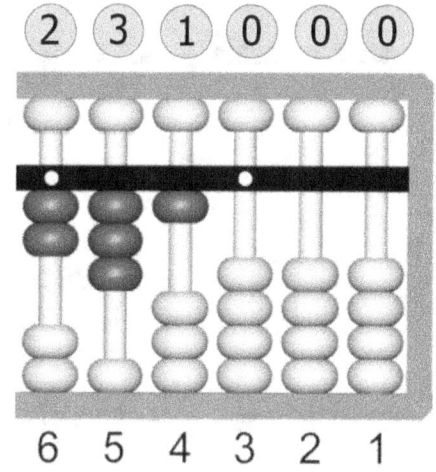

We will now **subtract 2064**

- -2000 • Column 6, unregister 2 lower beads
- • Column 5, do nothing
- -60 • Column 4, unregister 1 upper and 1 lower bead
- -4 • Column 3, unregister 4 lower beads

The abacus result is 2310

56 **Some examples of what to imagine step-by-step**

① Subtract 9 - 4

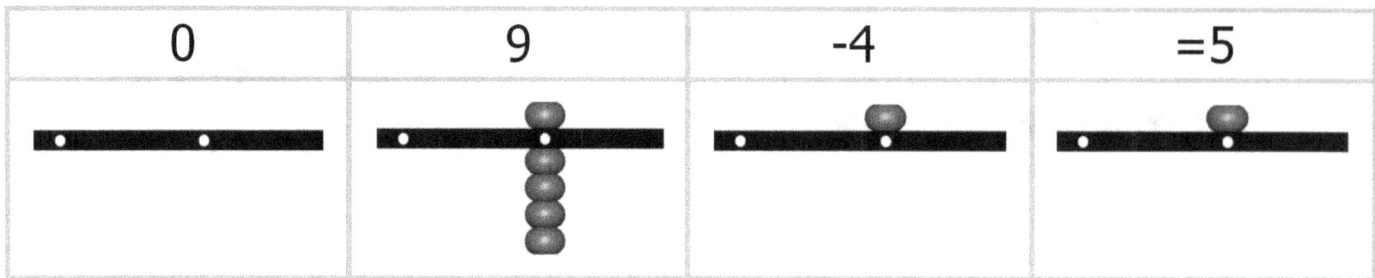

② Subtract 38 - 3

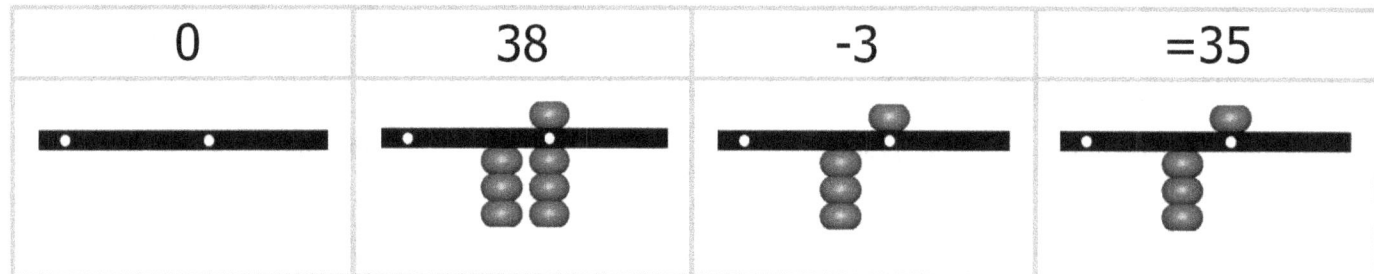

③ Subtract 78 - 6

| 0 | 78 | -6 | =72 |

④ Subtract 95 - 15

| 0 | 95 | -15 | =80 |

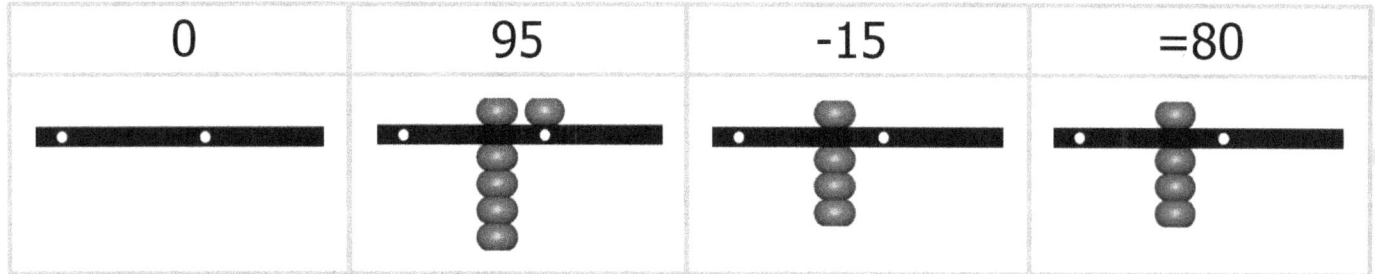

⑤ Subtract 87 - 17

| 0 | 87 | -17 | =70 |

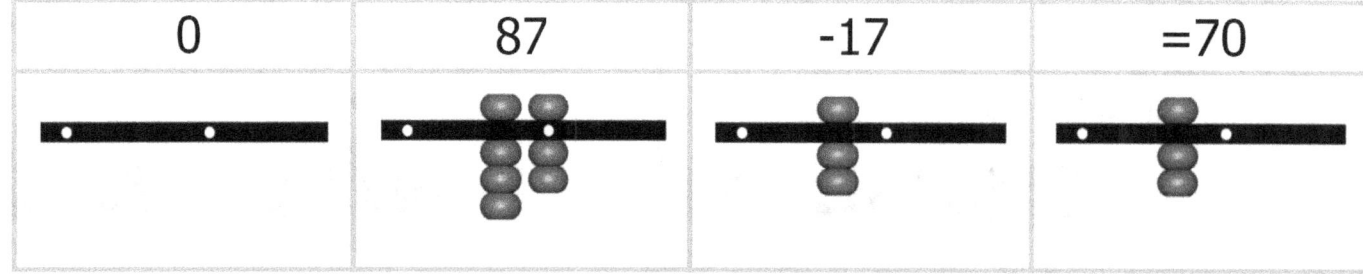

Some examples of what to imagine step-by-step

6) Subtract 49 - 15

0	49	-15	=34

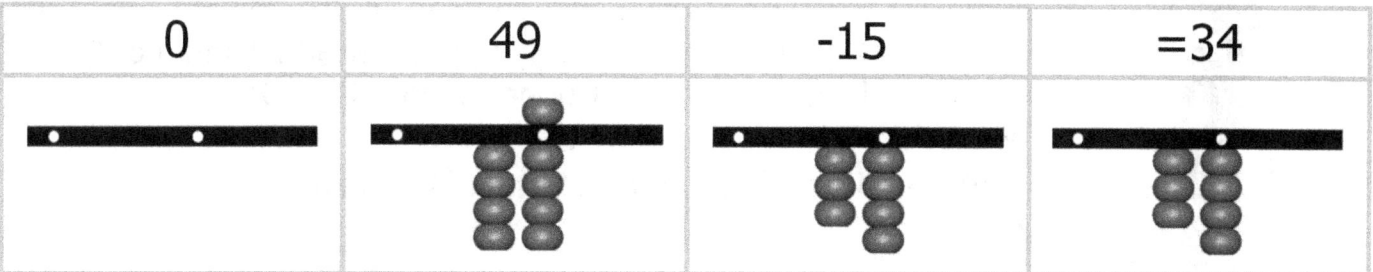

7) Subtract 18 - 14

0	18	-14	=4

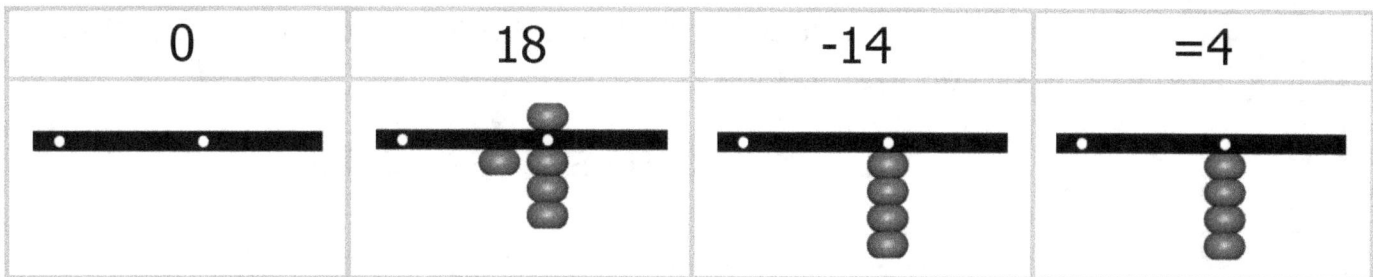

8) Subtract 94 - 63

0	94	-63	=31

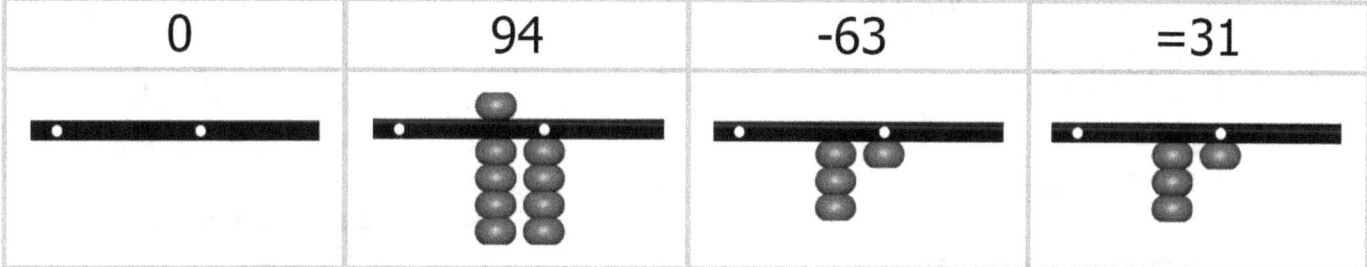

9) Subtract 95 - 62

0	95	-62	=33

Time to use the **workbook**! Go to workbook **page 43**.

Not enough beads in the column for the subtraction — Part 8

When you don't have enough beads, move to the next LEFT column to help

For example, when you try to subtract 8 from an already registered number 12, you don't have enough beads in the column where the 2 of the 12 is, to do it. You can only unregister a maximum of 9 in each column (4 lower beads and 1 upper bead, -4-5=-9).
When this happens, we need to use the
'Not enough beads list for subtraction'.

-1=-10+9
-2=-10+8
-3=-10+7
-4=-10+6
-5=-10+5
-6=-10+4
-7=-10+3
-8=-10+2
-9=-10+1

How to use the 'Not enough beads list for subtraction'

Let's say we need to subtract 8 from a column but we don't have enough beads.

Look at the list, **-8=-10+2**

10 is the number to **unregister**, in the next **LEFT** column (1 lower bead).

2 is the number to **register** in our column.

Example: 12 - 8

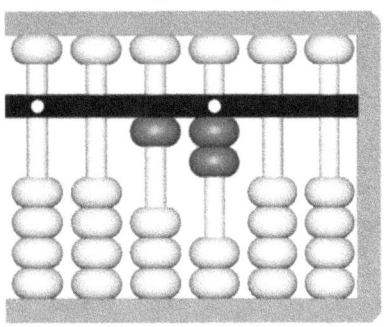

0 0 1 2 0 0

6 5 4 3 2 1
We will register 12

10 • Column 4, register 1 lower bead
2 • Column 3, register 2 lower beads

Unregister means move away from the beam (subtract)!

The abacus reads 12

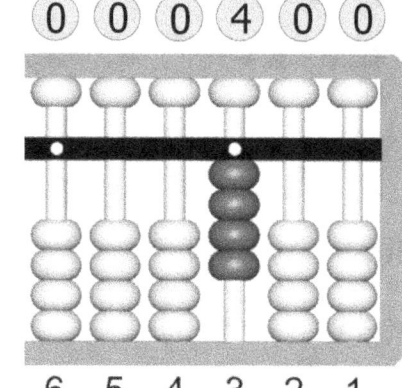

0 0 0 4 0 0

6 5 4 3 2 1
We will now subtract 8
There are not enough beads in column 3 to subtract 8, move to the next LEFT column to help. First we must think **-8=-10+2** see the
'Not enough beads list for subtraction'

-10 • Column 4, **unregister** 1 lower bead to subtract 10
+2 • Column 3, **register** 2 lower beads to add 2 (-8=-10+2)

The abacus result is 4

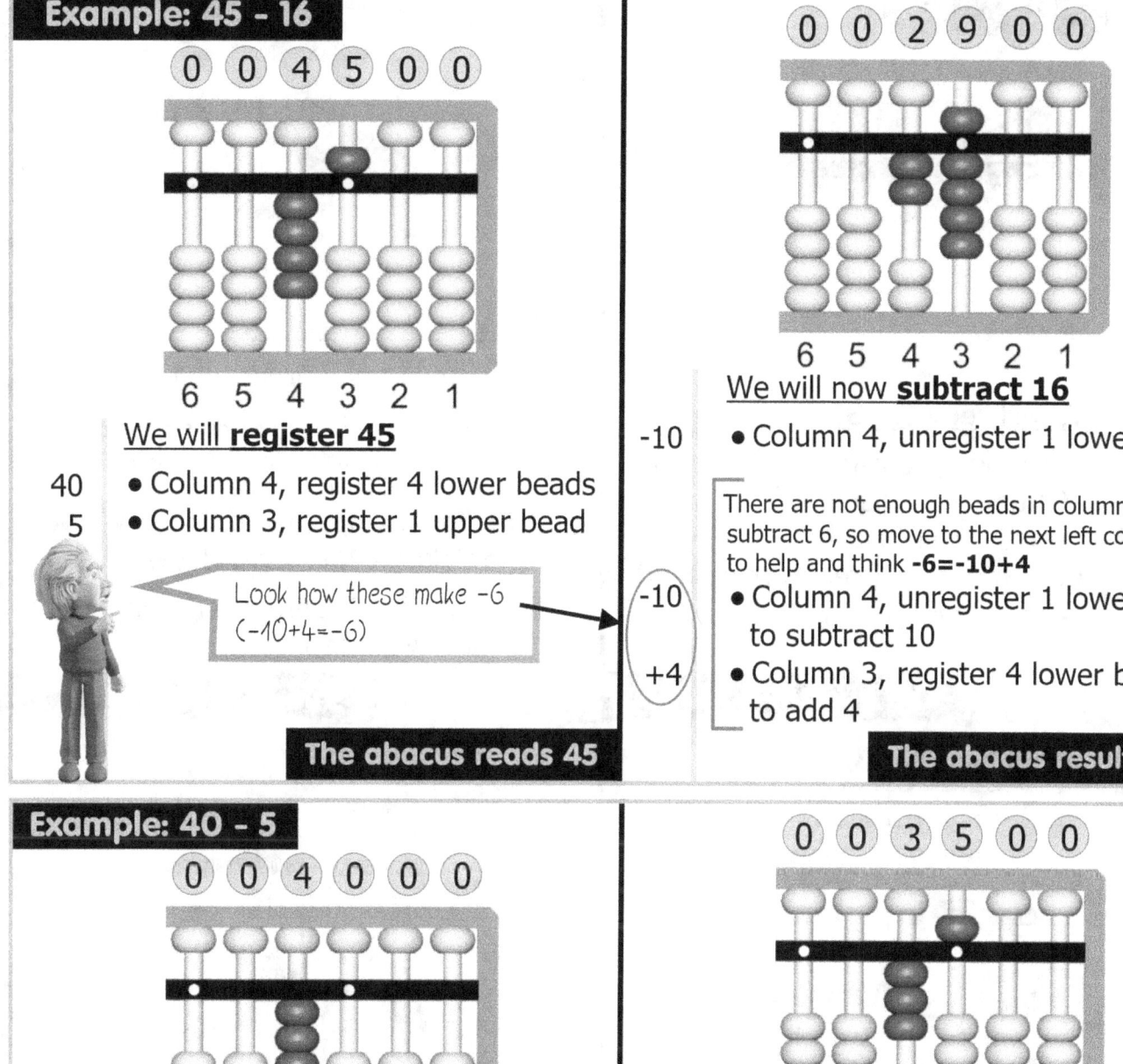

More subtraction examples (when we don't have enough beads)

Example: 377 - 186

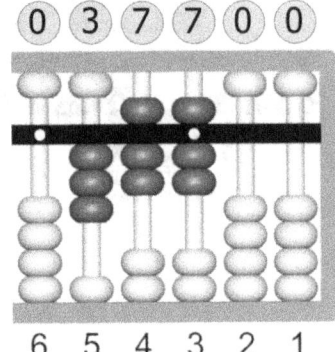

We will **register 377**

- 300 • Column 5, register 3 lower beads
- 70 • Column 4, register 1 upper bead and 2 lower beads
- 7 • Column 3, register 1 upper bead and 2 lower beads

The abacus reads 377

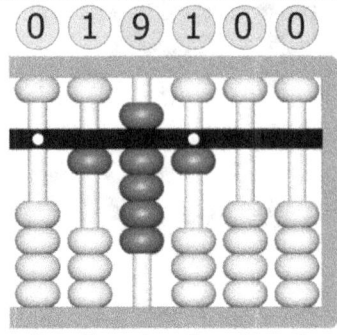

We will now **subtract 186**

- -100 • Column 5, unregister 1 lower bead

There are not enough beads in column 4 to subtract 8, so move to the next left column to help and think **-8=-10+2**
- -100 • Column 5, unregister 1 lower bead
- +20 • Column 4, register 2 lower beads

- -6 • Column 3, unregister 1 upper and 1 lower bead

The abacus result is 191

Example: 663 - 586

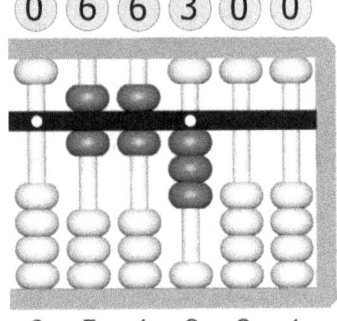

We will **register 663**

- 600 • Column 5, register 1 upper bead and 1 lower bead
- 60 • Column 4, register 1 upper bead and 1 lower bead
- 3 • Column 3, register 3 lower beads

Look how these make -80 and these make -6

The abacus reads 663

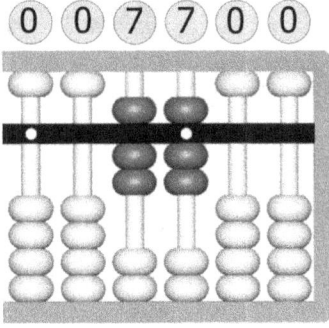

We will **subtract 586**

- -500 • Column 5, unregister 1 upper bead

There are not enough beads in column 4 to subtract 8, so move to the next left column to help and think **-8=-10+2**
- -100 • Column 5, unregister 1 lower bead
- +20 • Column 4, register 2 lower beads

There are not enough beads in column 3 to subtract 6, so move to the next left column to help and think **-6=-10+4**
- -10 • Column 4, unregister 1 lower bead
- +4 • Column 3, register 1 upper bead and unregister 1 lower bead (+5-1=+4)

The abacus result is 77

More subtraction examples (when we don't have enough beads)

Example: 8544 - 600

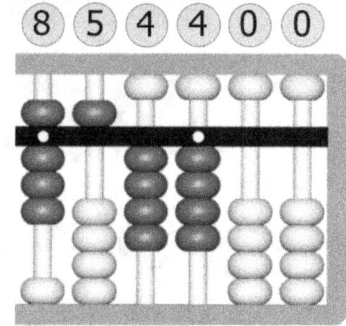

6 5 4 3 2 1

We will **register 8544**

- 8000 • Column 6, register 1 upper and 3 lower beads
- 500 • Column 5, register 1 upper bead
- 40 • Column 4, register 4 lower beads
- 4 • Column 3, register 4 lower beads

The abacus reads 8544

We will now **subtract 600**

There are not enough beads in column 5 to subtract 6, so move to the next left column to help and think **-6=-10+4**

- −1000 • Column 6, unregister 1 lower bead
- +400 • Column 5, register 4 lower beads
- • Columns 4 & 3, do nothing

The abacus result is 7944

Example: 9813 - 432

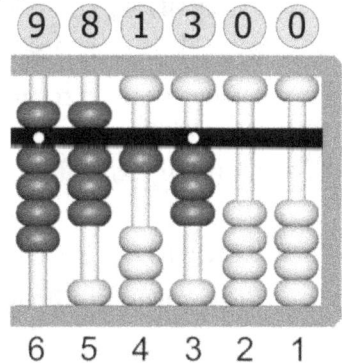

6 5 4 3 2 1

We will **register 9813**

- 9000 • Column 6, register 1 upper and 4 lower beads
- 800 • Column 5, register 1 upper and 3 lower beads
- 10 • Column 4, register 1 lower bead
- 3 • Column 3, register 3 lower beads

The abacus reads 9813

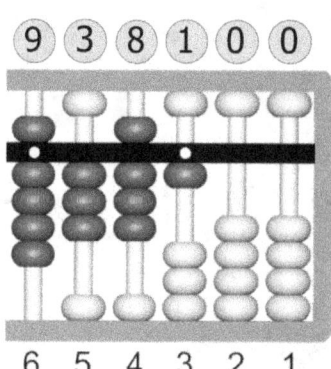

6 5 4 3 2 1

We will now **subtract 432**

- −400 • Column 5, unregister 1 upper bead and register 1 lower bead

Not enough beads in column 4 to subtract 3, so think **-3=-10+7**

- −100 • Column 5, unregister 1 lower bead
- +50 • Column 4, register 1 upper bead
- +20 • Column 4, register 2 lower beads
- −2 • Column 3, unregister 2 lower beads

The abacus result is 9381

62 More subtraction examples (when we don't have enough beads)

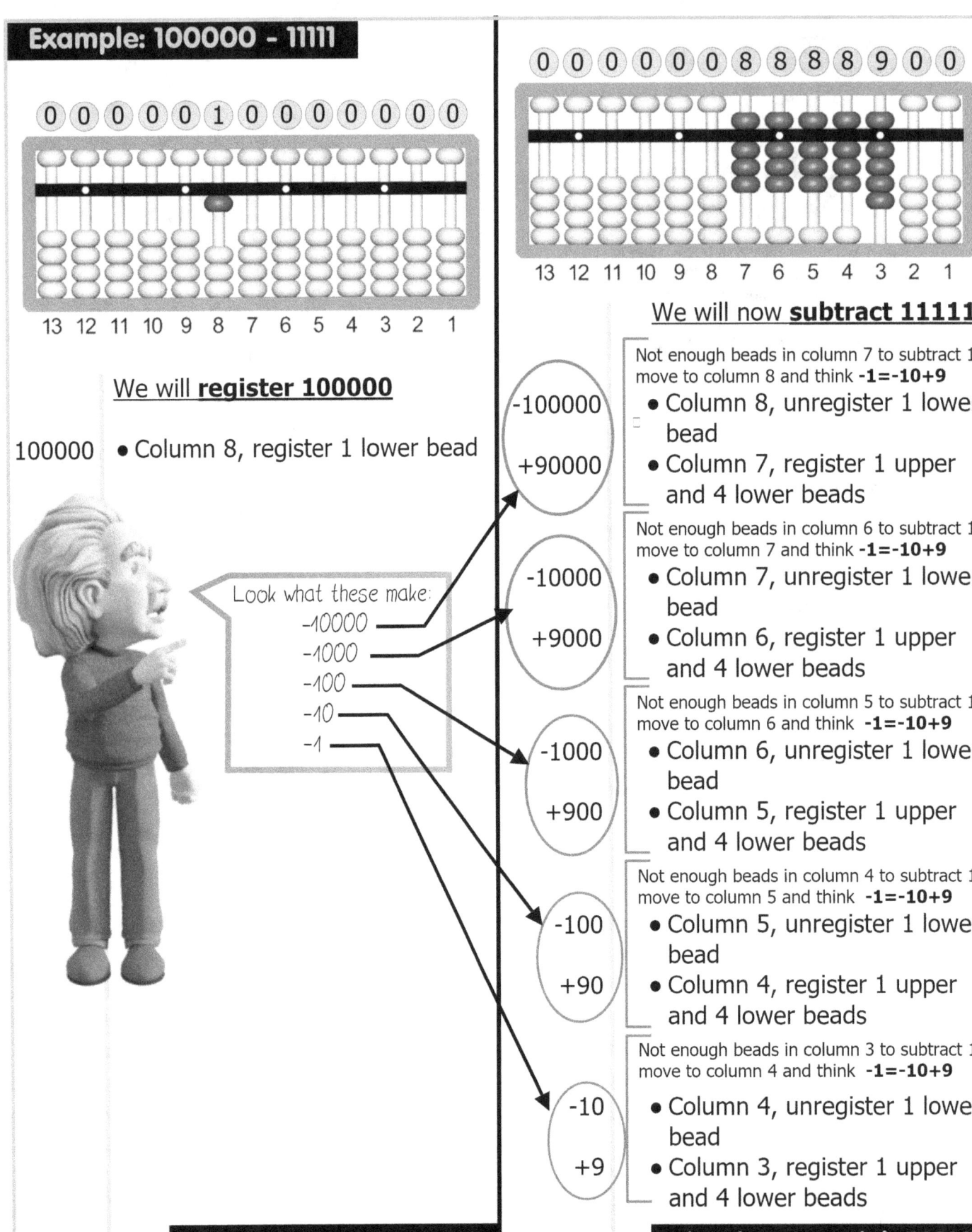

Some examples of what to imagine step-by-step

① Subtract 14 - 5

0	14	-5	=9

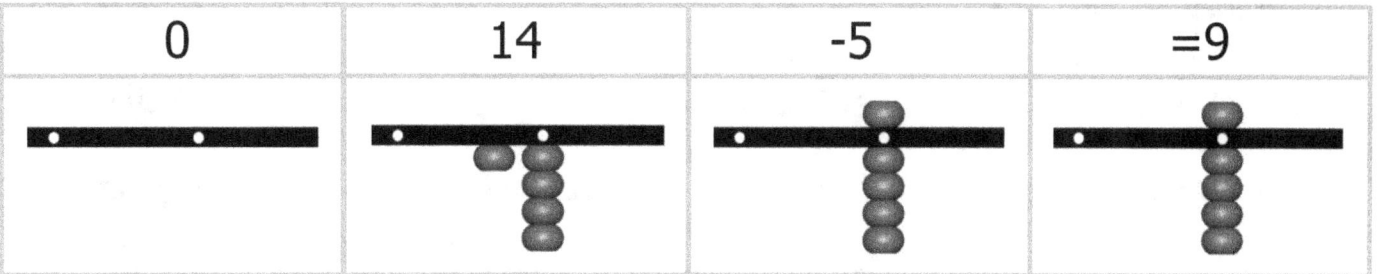

② Subtract 17 - 8

0	17	-8	=9

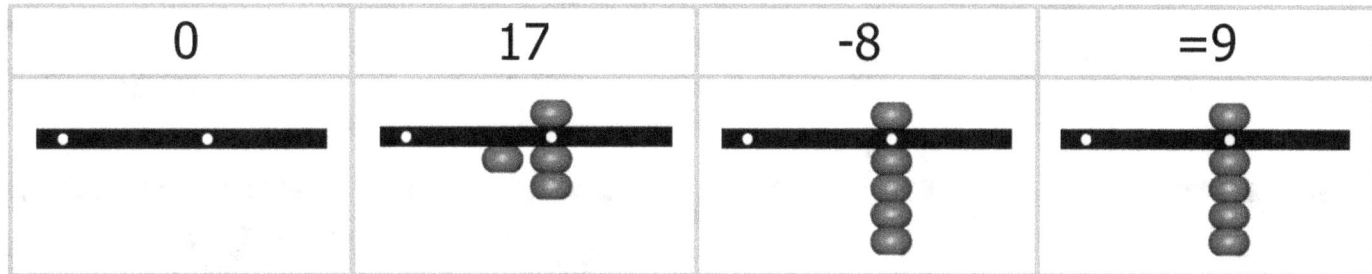

③ Subtract 63 - 7

0	63	-7	=56

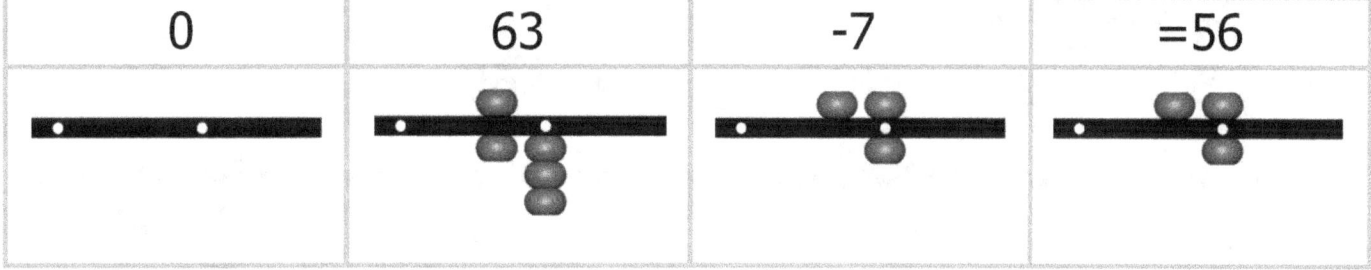

④ Subtract 95 - 16

0	95	-16	=79

⑤ Subtract 76 - 18

0	76	-18	=58

Some examples of what to imagine step-by-step

⑥ Subtract 45 - 7

0	45	-7	=38

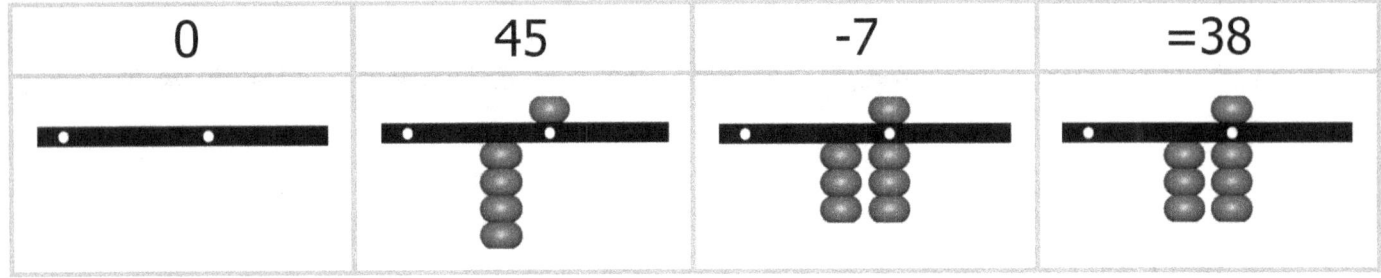

⑦ Subtract 455 - 65

0	455	-65	=390

⑧ Subtract 872 - 65

0	872	-65	=807

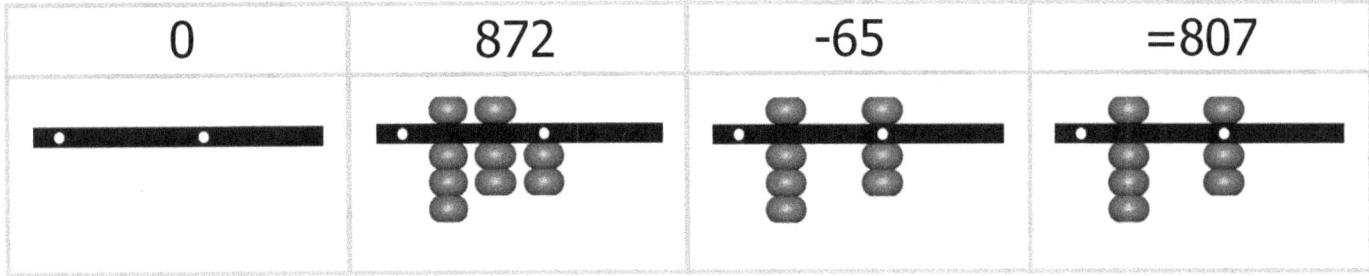

⑨ Subtract 895 - 66

0	895	-66	=829

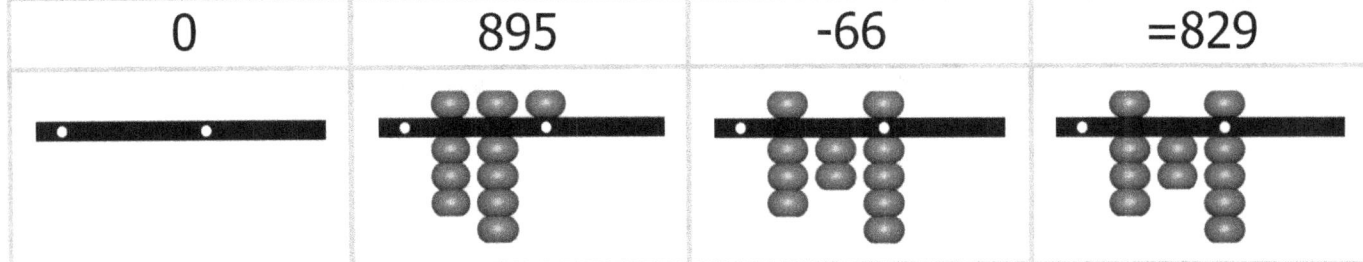

Time to use the **workbook!** Go to workbook **page 50**.

Skipped columns when subtracting   Part 9

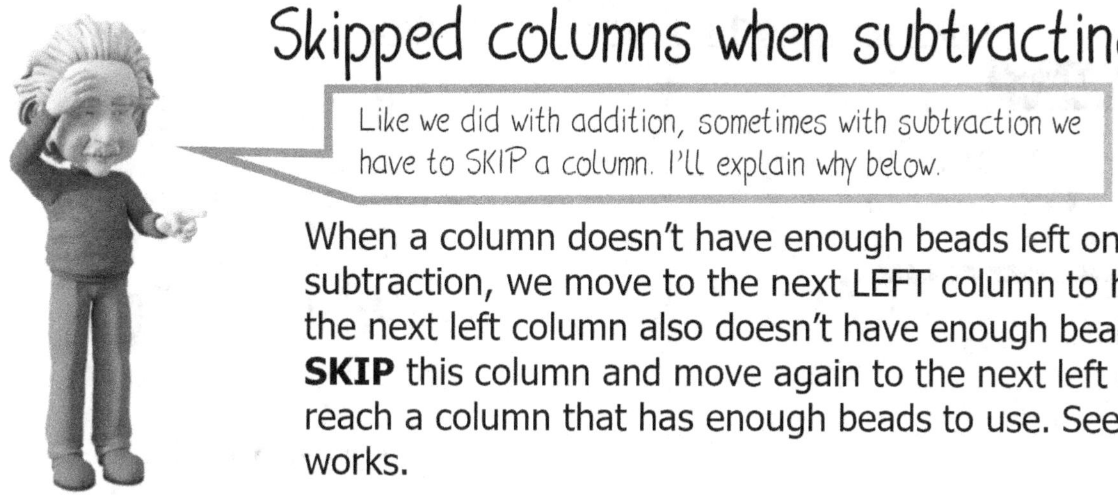

Like we did with addition, sometimes with subtraction we have to SKIP a column. I'll explain why below.

When a column doesn't have enough beads left on it to make the subtraction, we move to the next LEFT column to help. Sometimes the next left column also doesn't have enough beads on it, so we **SKIP** this column and move again to the next left column until you reach a column that has enough beads to use. See below how it works.

Important!

- We will **SKIP** a column when there are not enough beads to use in that column.
- We will see this symbol when we need to **SKIP** ⬅ a column (move on to the next left column).
- We will see this symbol when we need to **MOVE BACK** ➡ a column (move on to the next right column).
- We will **REGISTER** all beads in any skipped columns (with addition we unregistered, here we do the opposite).

Example: 300 - 5

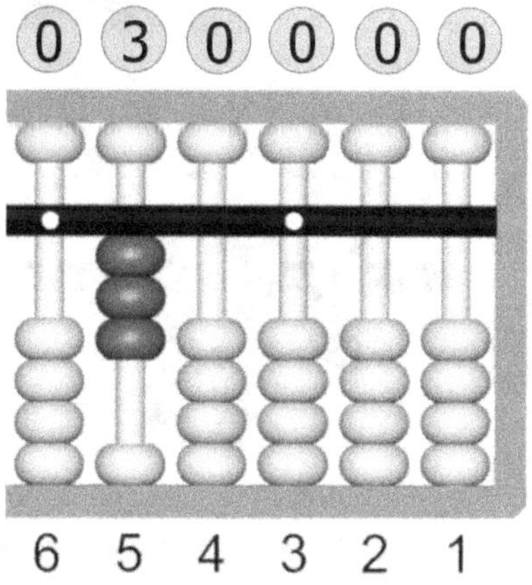

0 3 0 0 0 0

6 5 4 3 2 1
We will register 300

300 • Column 5, register 3 lower beads

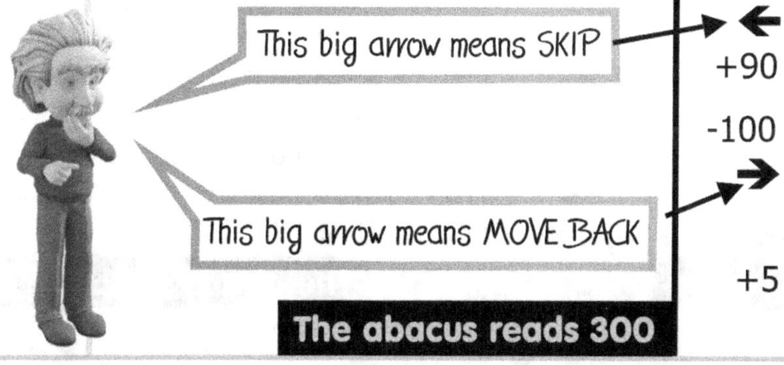

This big arrow means SKIP

This big arrow means MOVE BACK

The abacus reads 300

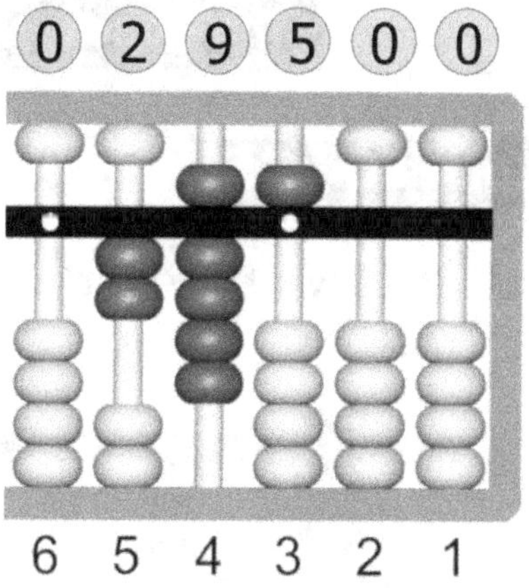

0 2 9 5 0 0

6 5 4 3 2 1
We will now subtract 5

There are not enough beads in column 3 to subtract 5, move to column 4, think **-5=-10+5**

⬅
+90 • Column 4, **SKIP** this column and register all beads

-100 • Column 5, unregister 1 lower bead

➡ • **MOVE BACK** past the skipped column 4

+5 • Column 3, register 1 upper bead

The abacus result is 295

More skipped column examples

Example: 1000 - 1

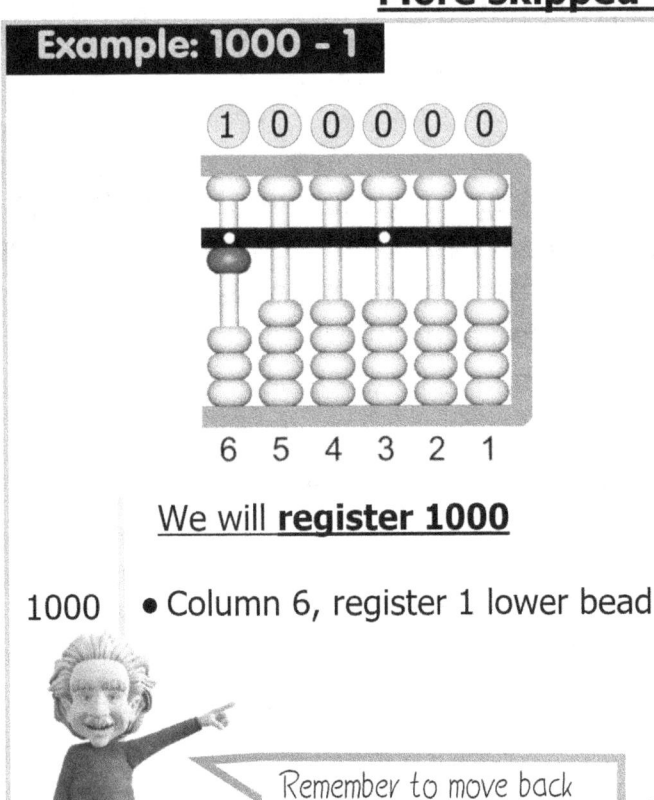

We will **register 1000**

1000 • Column 6, register 1 lower bead

Remember to move back past the skipped columns!

The abacus reads 1000

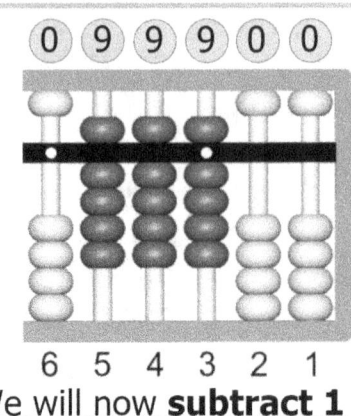

We will now **subtract 1**

There are not enough beads in column 3 to unregister 1, move to column 4, think **-1=-10+9**

← +90 • Column 4, **SKIP** this column and register all beads
← +900 • Column 5, **SKIP** this column and register all beads
-1000 • Column 6, unregister 1 lower bead
→ → • **MOVE BACK** past the skipped columns 5 and 4
+9 • Column 3, register 1 upper and 4 lower beads

The abacus result is 999

Example: 804 - 5

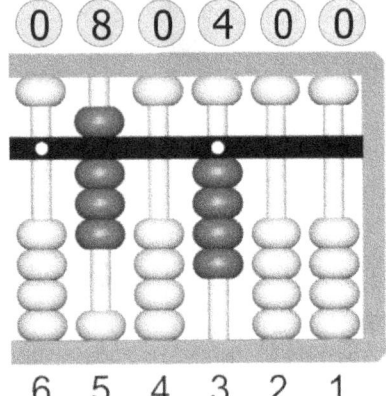

We will **register 804**

800 • Column 5, register 1 upper and 3 lower beads
• Column 4, do nothing
4 • Column 3, register 4 lower beads

The abacus reads 804

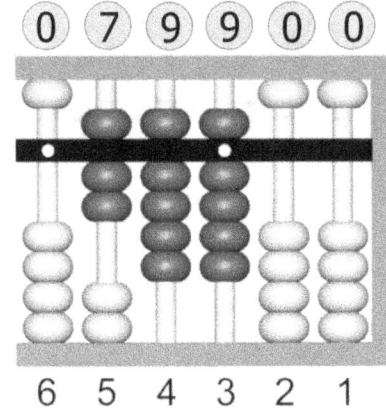

We will now **subtract 5**

There are not enough beads in column 3 to unregister 5, move to column 4, think **-5=-10+5**

← +90 • Column 4, **SKIP** this column and register all beads
-100 • Column 5, unregister 1 lower bead
→ • **MOVE BACK** past the skipped column 4
+5 • Column 3, register 1 upper bead

The abacus result is 799

Some examples of what to imagine step-by-step

① Subtract 604 - 5

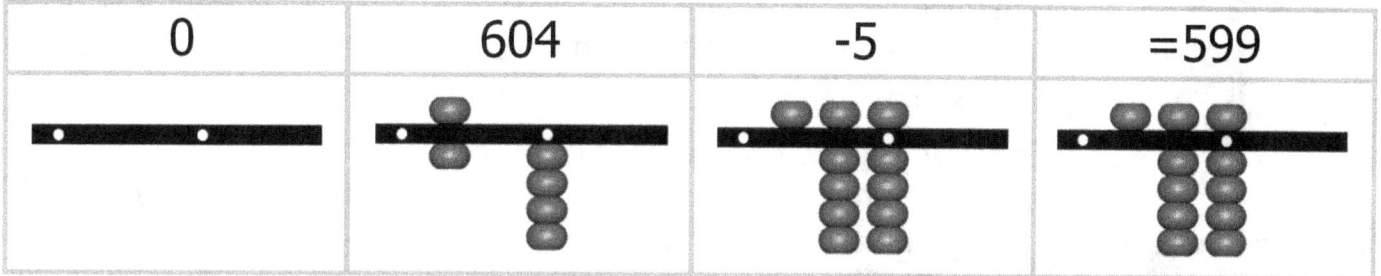

② Subtract 1000 - 8

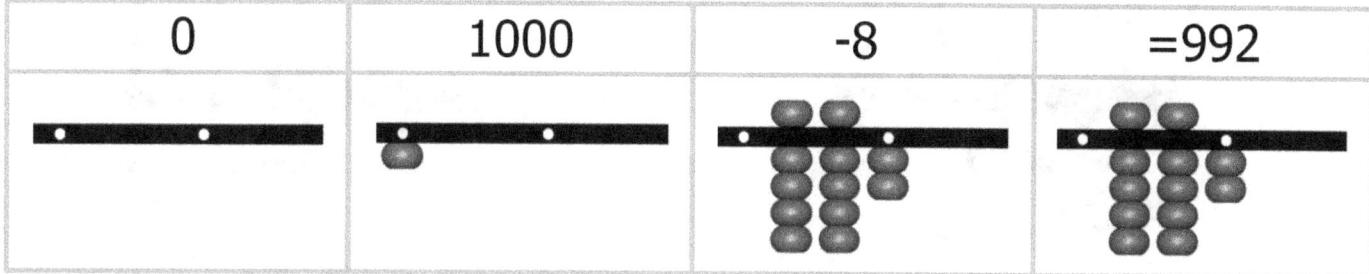

③ Subtract 306 - 9

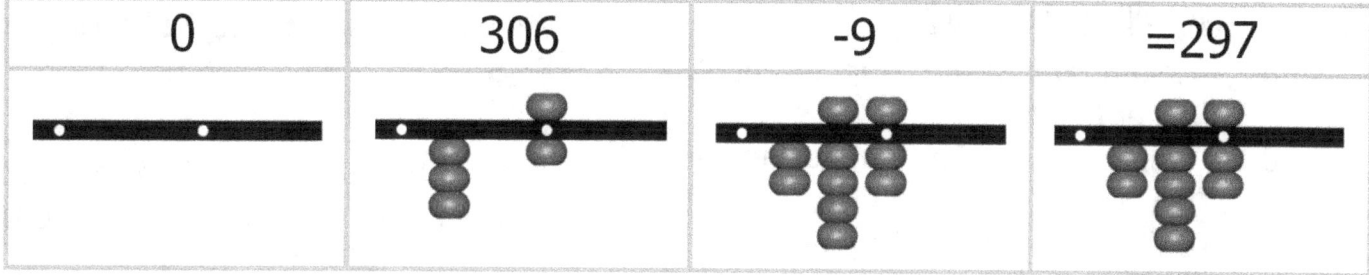

④ Subtract 301 - 2

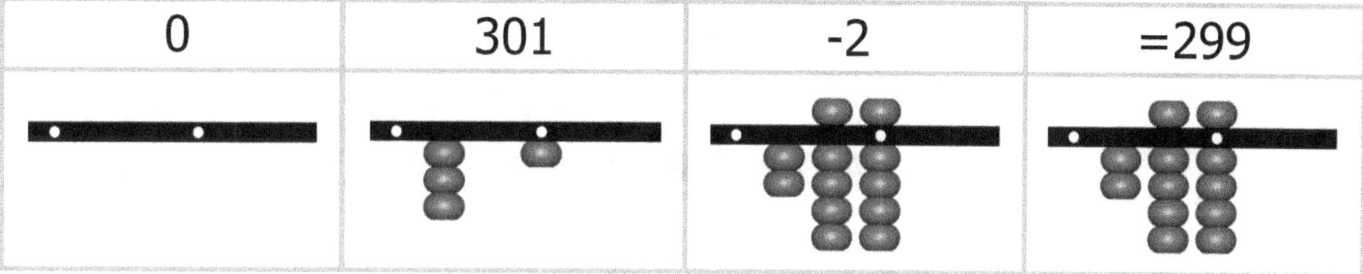

Time to use the **workbook!** Go to workbook **page 53**.

Subtraction of 3 or more numbers   Part 10

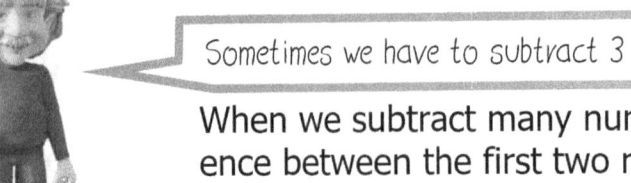

Sometimes we have to subtract 3 or more numbers, here's how.

When we subtract many numbers on the abacus, just find the difference between the first two numbers, then subtract the next number to get the new difference.

Keep subtracting one number from the difference of the previous numbers until all the numbers have been subtracted.

Example: 998 - 221 - 125

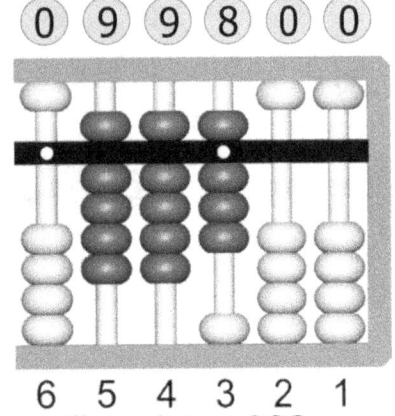

We will register 998

- 900 • Column 5, register 1 upper and 4 lower beads
- 90 • Column 4, register 1 upper and 4 lower beads
- 8 • Column 3, register 1 upper and 3 lower beads

The abacus reads 998

We will now subtract 221 from 998

- -200 • Column 5, unregister 2 lower beads
- -20 • Column 4, unregister 2 lower beads
- -1 • Column 3, unregister 1 lower beads

The abacus now displays 777

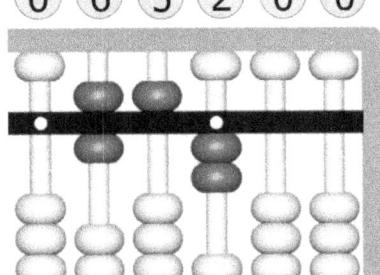

We will now subtract 125 from 777

- -100 • Column 5, unregister 1 lower bead
- -20 • Column 4, unregister 2 lower beads
- -5 • Column 3, unregister 1 upper bead

The abacus result is 652

Subtraction of 3 or more numbers

Example: 644662 - 522330 - 2140

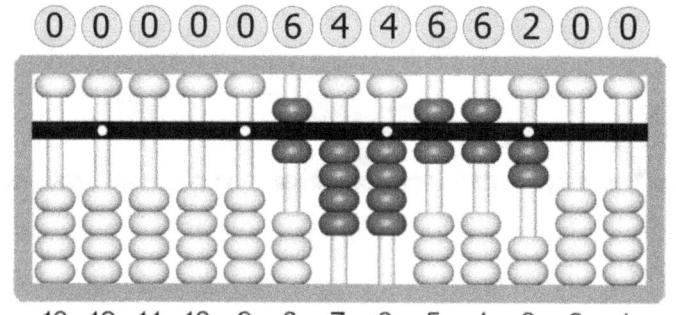

We will **register 644662**

600000	• Column 8, register 1 upper and 1 lower bead
40000	• Column 7, register 4 lower beads
4000	• Column 6, register 4 lower beads
600	• Column 5, register 1 upper and 1 lower bead
60	• Column 4, register 1 upper and 1 lower bead
2	• Column 3, register 2 lower beads

The abacus reads 644662

We will now **subtract 522330 from 644662**

-500000	• Column 8, unregister 1 upper bead
-20000	• Column 7, unregister 2 lower beads
-2000	• Column 6, unregister 2 lower beads
-300	• Column 5, unregister 1 upper bead and register 2 lower beads
-30	• Column 4, unregister 1 upper bead and register 2 lower beads
	• Column 3, do nothing

The abacus now displays 122332

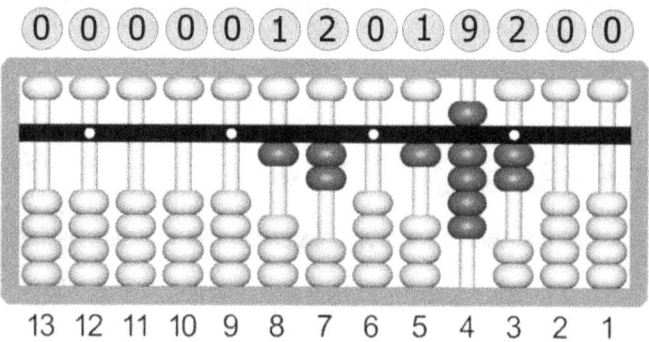

We will now **subtract 2140 from 122332**

-2000	• Column 6, unregister 2 lower beads
-100	• Column 5, unregister 1 lower bead

There are not enough beads in column 4 to unregister 4 more, move to column 5, think **-4=-10+6**

-100	• Column 5, unregister 1 lower bead
+60	• Column 4, register 1 upper and 1 lower bead
	• Column 3, do nothing

The abacus result is 120192

Some examples of what to imagine step-by-step

① Subtract 14 - 5 - 4

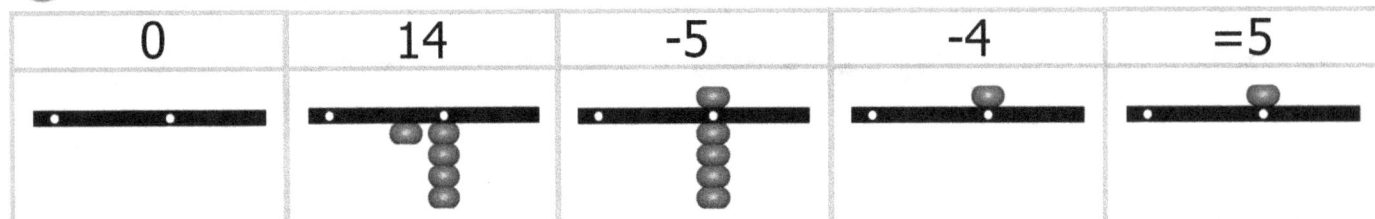

| 0 | 14 | -5 | -4 | =5 |

② Subtract 35 - 15 - 5

| 0 | 35 | -15 | -5 | =15 |

③ Subtract 87 - 27 - 20

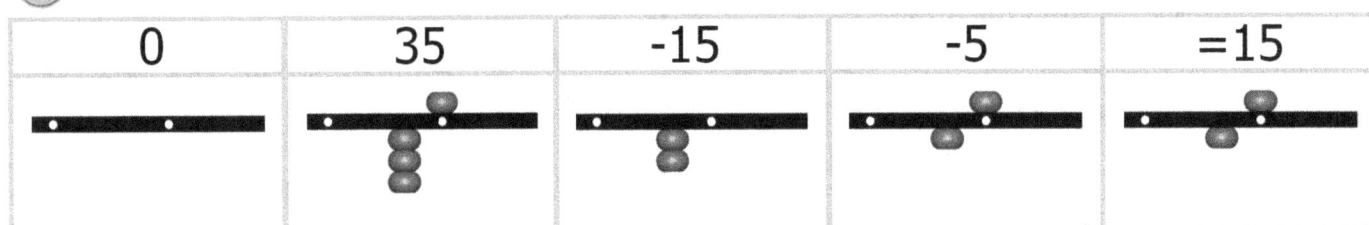

| 0 | 87 | -27 | -20 | =40 |

④ Subtract 205 - 15 - 4

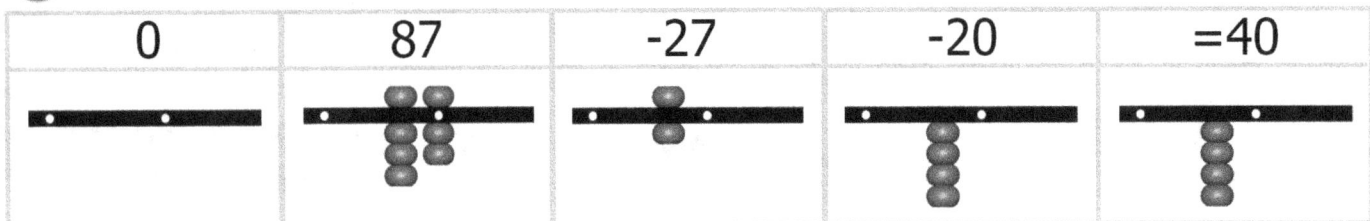

| 0 | 205 | -15 | -4 | =186 |

⑤ Subtract 74 - 3 - 31 - 20

| 0 | 74 | -3 | -31 | -20 | =20 |

Addition and Subtraction together

Now we will add and subtract numbers in the same calculation, here's how.

We just need to use a combination of adding and subtracting in the same way that we have already learnt to do.

Here are some examples:

Example: 84 - 63 + 11

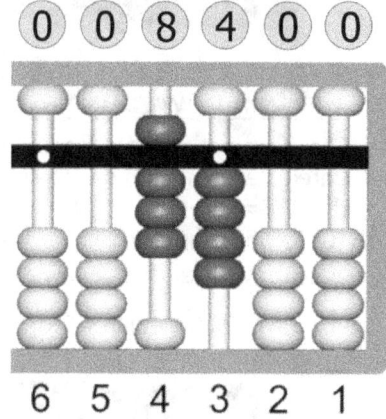

We will register 84

- 80 • Column 4, register 1 upper and 3 lower beads
- 4 • Column 3, register 4 lower beads

The abacus reads 84

We will now subtract 63 from 84

- -60 • Column 4, unregister 1 upper bead register 1 lower bead
- -3 • Column 3, unregister 3 lower beads

The abacus now displays 21

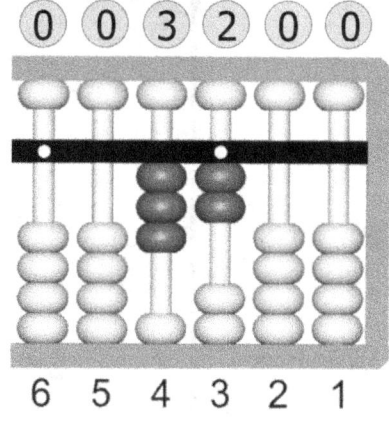

We will now add 11 to 21

- +10 • Column 4, register 1 lower bead
- +1 • Column 3, register 1 lower bead

The abacus result is 32

More addition and subtraction examples

Example: 405 - 82 + 61 - 83

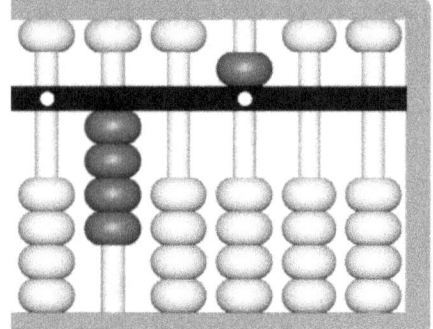

We will register 405

- 400 • Column 5, register 4 lower beads
- • Column 4, do nothing
- 5 • Column 3, register 1 upper bead

The abacus reads 405

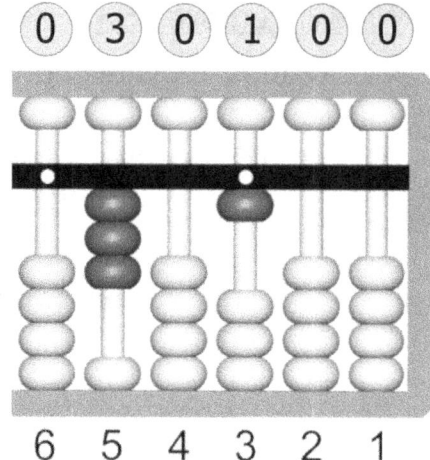

We will now subtract 82

There are not enough beads in column 4 to unregister 8 more, move to column 5, think **-8=-10+2**

- -100 • Column 5, unregister 1 lower bead
- +20 • Column 4, register 2 lower beads
- -2 • Column 3, unregister 1 upper bead and register 3 lower beads

The abacus result is 323

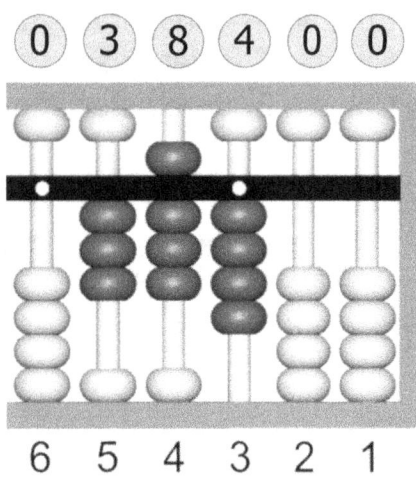

We will now add 61

- 60 • Column 4, register 1 upper and 1 lower bead
- 1 • Column 3, register 1 lower bead

The abacus reads 384

We will now subtract 83

- -80 • Column 4, unregister 1 upper and 3 lower beads
- -3 • Column 3, unregister 3 lower beads

The abacus result is 301

More addition and subtraction examples

73

Example: 577 + 421 - 263 + 153

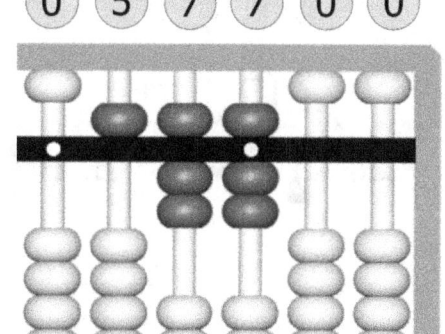

We will **register 577**

- 500 • Column 5, register 1 upper bead
- 70 • Column 4, register 1 upper and 2 lower beads
- 7 • Column 3, register 1 upper bead and 2 lower beads

The abacus reads 577

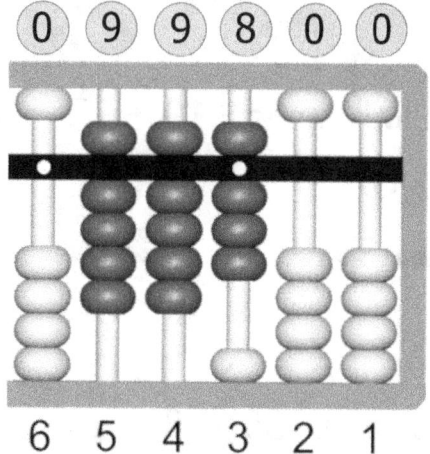

We will now **add 421**

- +400 • Column 5, register 4 lower beads
- +20 • Column 4, register 2 lower beads
- +1 • Column 3, register 1 lower bead

The abacus result is 998

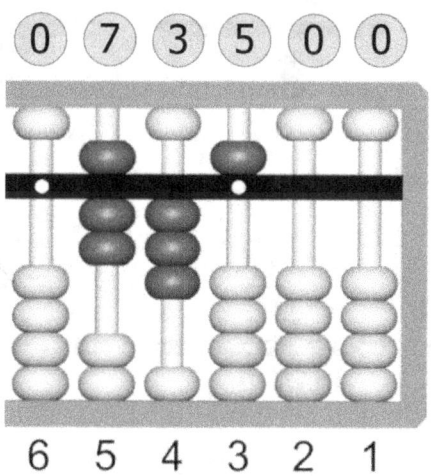

We will now **subtract 263**

- -200 • Column 5, unregister 2 lower beads
- -60 • Column 4, unregister 1 upper and 1 lower bead
- -3 • Column 3, unregister 3 lower beads

The abacus reads 735

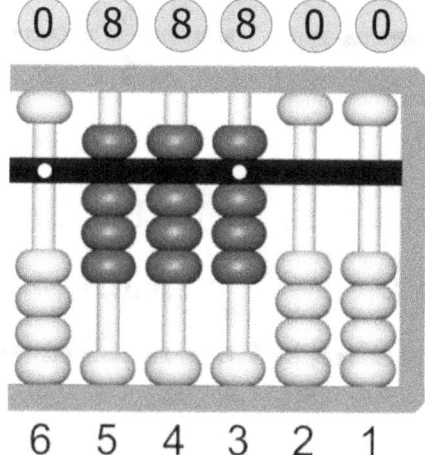

We will now **add 153**

- +100 • Column 5, register 1 lower bead
- +50 • Column 4, register 1 upper bead
- +3 • Column 3, register 3 lower beads

The abacus result is 888

74 **Some examples of what to imagine step-by-step**

① Calculate 65 - 10 + 12

0	65	-10	+12	=67

② Calculate 35 - 12 + 18

0	35	-12	+18	=41

③ Calculate 75 + 15 - 60

0	75	+15	-60	=30

④ Calculate 205 + 81 - 16

0	205	+81	-16	=270

⑤ Calculate 90 - 30 - 15 + 18

0	90	-30	-15	+18	=63

Time to use the **workbook**! Go to workbook **page 56**.

Using the reusable workbook pages   Part 11

In the last part of the workbook we will be using reusable work pages (see workbook pages 58 to 71). This means that they are meant to be done over-and-over until you are proficient at firstly the actual abacus and then finally the imaginary.

- Select a reusable workbook page of your choice
- Decide if you will make the calculations either with the abacus or with an imaginary abacus
- Select your column (A, B, C etc..)

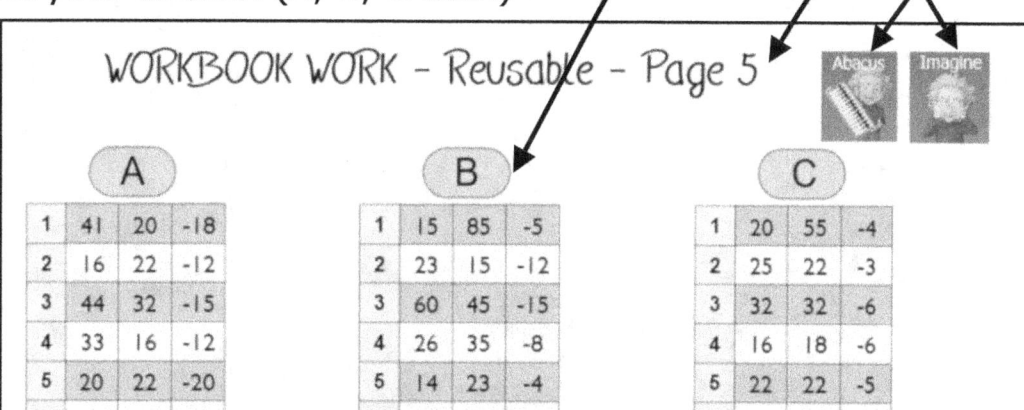

- Write down your Page / Column number on the answer sheet (see blank answer sheets pages 79 to 99 of the instruction book)
- Write down the answers in that column

Write the ANSWERS for your reusable work pages

Page / Column 5 / B	Page / Column	Page / Column	Page / Column	Page / Column
1 95	1	1	1	1
2 26	2	2	2	2
3 90	3	3	3	3
4 53	4	4	4	4
5 33	5	5	5	5

- Check your answers with the answer sheet (pages 109 to 122 of the workbook)

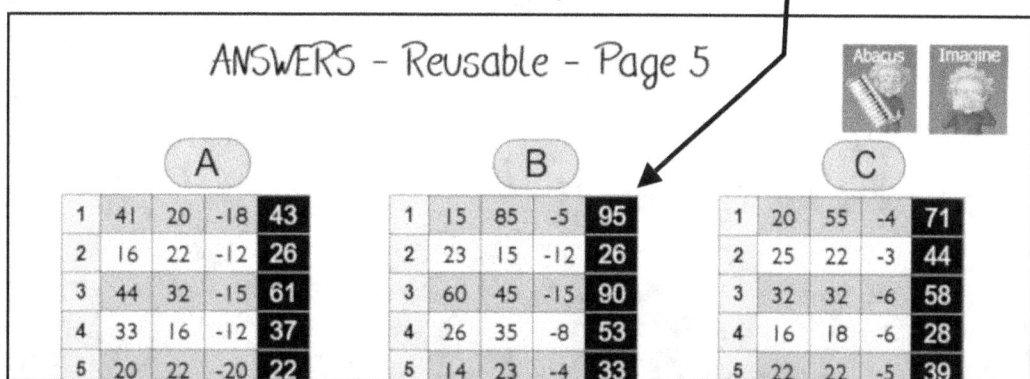

Notes

Blank sheets for reusable workbook page answers

Write the ANSWERS for your reusable work pages

Page / Column		Page / Column		Page / Column		Page / Column		Page / Column	
1		1		1		1		1	
2		2		2		2		2	
3		3		3		3		3	
4		4		4		4		4	
5		5		5		5		5	
6		6		6		6		6	
7		7		7		7		7	
8		8		8		8		8	
9		9		9		9		9	
10		10		10		10		10	
11		11		11		11		11	
12		12		12		12		12	
13		13		13		13		13	
14		14		14		14		14	
15		15		15		15		15	
16		16		16		16		16	
17		17		17		17		17	
18		18		18		18		18	
19		19		19		19		19	
20		20		20		20		20	
21		21		21		21		21	
22		22		22		22		22	
23		23		23		23		23	
24		24		24		24		24	
25		25		25		25		25	
26		26		26		26		26	
27		27		27		27		27	
28		28		28		28		28	
29		29		29		29		29	
30		30		30		30		30	

Write the ANSWERS for your reusable work pages

Page / Column		Page / Column		Page / Column		Page / Column		Page / Column	
1		1		1		1		1	
2		2		2		2		2	
3		3		3		3		3	
4		4		4		4		4	
5		5		5		5		5	
6		6		6		6		6	
7		7		7		7		7	
8		8		8		8		8	
9		9		9		9		9	
10		10		10		10		10	
11		11		11		11		11	
12		12		12		12		12	
13		13		13		13		13	
14		14		14		14		14	
15		15		15		15		15	
16		16		16		16		16	
17		17		17		17		17	
18		18		18		18		18	
19		19		19		19		19	
20		20		20		20		20	
21		21		21		21		21	
22		22		22		22		22	
23		23		23		23		23	
24		24		24		24		24	
25		25		25		25		25	
26		26		26		26		26	
27		27		27		27		27	
28		28		28		28		28	
29		29		29		29		29	
30		30		30		30		30	

Write the ANSWERS for your reusable work pages

Page / Column		Page / Column		Page / Column		Page / Column		Page / Column	
1		1		1		1		1	
2		2		2		2		2	
3		3		3		3		3	
4		4		4		4		4	
5		5		5		5		5	
6		6		6		6		6	
7		7		7		7		7	
8		8		8		8		8	
9		9		9		9		9	
10		10		10		10		10	
11		11		11		11		11	
12		12		12		12		12	
13		13		13		13		13	
14		14		14		14		14	
15		15		15		15		15	
16		16		16		16		16	
17		17		17		17		17	
18		18		18		18		18	
19		19		19		19		19	
20		20		20		20		20	
21		21		21		21		21	
22		22		22		22		22	
23		23		23		23		23	
24		24		24		24		24	
25		25		25		25		25	
26		26		26		26		26	
27		27		27		27		27	
28		28		28		28		28	
29		29		29		29		29	
30		30		30		30		30	

Write the ANSWERS for your reusable work pages

Page / Column		Page / Column		Page / Column		Page / Column		Page / Column	
1		1		1		1		1	
2		2		2		2		2	
3		3		3		3		3	
4		4		4		4		4	
5		5		5		5		5	
6		6		6		6		6	
7		7		7		7		7	
8		8		8		8		8	
9		9		9		9		9	
10		10		10		10		10	
11		11		11		11		11	
12		12		12		12		12	
13		13		13		13		13	
14		14		14		14		14	
15		15		15		15		15	
16		16		16		16		16	
17		17		17		17		17	
18		18		18		18		18	
19		19		19		19		19	
20		20		20		20		20	
21		21		21		21		21	
22		22		22		22		22	
23		23		23		23		23	
24		24		24		24		24	
25		25		25		25		25	
26		26		26		26		26	
27		27		27		27		27	
28		28		28		28		28	
29		29		29		29		29	
30		30		30		30		30	

Write the ANSWERS for your reusable work pages

Page / Column		Page / Column		Page / Column		Page / Column		Page / Column	
1		1		1		1		1	
2		2		2		2		2	
3		3		3		3		3	
4		4		4		4		4	
5		5		5		5		5	
6		6		6		6		6	
7		7		7		7		7	
8		8		8		8		8	
9		9		9		9		9	
10		10		10		10		10	
11		11		11		11		11	
12		12		12		12		12	
13		13		13		13		13	
14		14		14		14		14	
15		15		15		15		15	
16		16		16		16		16	
17		17		17		17		17	
18		18		18		18		18	
19		19		19		19		19	
20		20		20		20		20	
21		21		21		21		21	
22		22		22		22		22	
23		23		23		23		23	
24		24		24		24		24	
25		25		25		25		25	
26		26		26		26		26	
27		27		27		27		27	
28		28		28		28		28	
29		29		29		29		29	
30		30		30		30		30	

Write the ANSWERS for your reusable work pages

Page / Column		Page / Column		Page / Column		Page / Column		Page / Column	
1		1		1		1		1	
2		2		2		2		2	
3		3		3		3		3	
4		4		4		4		4	
5		5		5		5		5	
6		6		6		6		6	
7		7		7		7		7	
8		8		8		8		8	
9		9		9		9		9	
10		10		10		10		10	
11		11		11		11		11	
12		12		12		12		12	
13		13		13		13		13	
14		14		14		14		14	
15		15		15		15		15	
16		16		16		16		16	
17		17		17		17		17	
18		18		18		18		18	
19		19		19		19		19	
20		20		20		20		20	
21		21		21		21		21	
22		22		22		22		22	
23		23		23		23		23	
24		24		24		24		24	
25		25		25		25		25	
26		26		26		26		26	
27		27		27		27		27	
28		28		28		28		28	
29		29		29		29		29	
30		30		30		30		30	

Write the ANSWERS for your reusable work pages

Page / Column		Page / Column		Page / Column		Page / Column		Page / Column	
1		1		1		1		1	
2		2		2		2		2	
3		3		3		3		3	
4		4		4		4		4	
5		5		5		5		5	
6		6		6		6		6	
7		7		7		7		7	
8		8		8		8		8	
9		9		9		9		9	
10		10		10		10		10	
11		11		11		11		11	
12		12		12		12		12	
13		13		13		13		13	
14		14		14		14		14	
15		15		15		15		15	
16		16		16		16		16	
17		17		17		17		17	
18		18		18		18		18	
19		19		19		19		19	
20		20		20		20		20	
21		21		21		21		21	
22		22		22		22		22	
23		23		23		23		23	
24		24		24		24		24	
25		25		25		25		25	
26		26		26		26		26	
27		27		27		27		27	
28		28		28		28		28	
29		29		29		29		29	
30		30		30		30		30	

Write the ANSWERS for your reusable work pages

Page / Column		Page / Column		Page / Column		Page / Column		Page / Column	
1		1		1		1		1	
2		2		2		2		2	
3		3		3		3		3	
4		4		4		4		4	
5		5		5		5		5	
6		6		6		6		6	
7		7		7		7		7	
8		8		8		8		8	
9		9		9		9		9	
10		10		10		10		10	
11		11		11		11		11	
12		12		12		12		12	
13		13		13		13		13	
14		14		14		14		14	
15		15		15		15		15	
16		16		16		16		16	
17		17		17		17		17	
18		18		18		18		18	
19		19		19		19		19	
20		20		20		20		20	
21		21		21		21		21	
22		22		22		22		22	
23		23		23		23		23	
24		24		24		24		24	
25		25		25		25		25	
26		26		26		26		26	
27		27		27		27		27	
28		28		28		28		28	
29		29		29		29		29	
30		30		30		30		30	

Write the ANSWERS for your reusable work pages

Page / Column		Page / Column		Page / Column		Page / Column		Page / Column	
1		1		1		1		1	
2		2		2		2		2	
3		3		3		3		3	
4		4		4		4		4	
5		5		5		5		5	
6		6		6		6		6	
7		7		7		7		7	
8		8		8		8		8	
9		9		9		9		9	
10		10		10		10		10	
11		11		11		11		11	
12		12		12		12		12	
13		13		13		13		13	
14		14		14		14		14	
15		15		15		15		15	
16		16		16		16		16	
17		17		17		17		17	
18		18		18		18		18	
19		19		19		19		19	
20		20		20		20		20	
21		21		21		21		21	
22		22		22		22		22	
23		23		23		23		23	
24		24		24		24		24	
25		25		25		25		25	
26		26		26		26		26	
27		27		27		27		27	
28		28		28		28		28	
29		29		29		29		29	
30		30		30		30		30	

Write the ANSWERS for your reusable work pages

Page / Column		Page / Column		Page / Column		Page / Column		Page / Column	
1		1		1		1		1	
2		2		2		2		2	
3		3		3		3		3	
4		4		4		4		4	
5		5		5		5		5	
6		6		6		6		6	
7		7		7		7		7	
8		8		8		8		8	
9		9		9		9		9	
10		10		10		10		10	
11		11		11		11		11	
12		12		12		12		12	
13		13		13		13		13	
14		14		14		14		14	
15		15		15		15		15	
16		16		16		16		16	
17		17		17		17		17	
18		18		18		18		18	
19		19		19		19		19	
20		20		20		20		20	
21		21		21		21		21	
22		22		22		22		22	
23		23		23		23		23	
24		24		24		24		24	
25		25		25		25		25	
26		26		26		26		26	
27		27		27		27		27	
28		28		28		28		28	
29		29		29		29		29	
30		30		30		30		30	

Write the ANSWERS for your reusable work pages

Page / Column		Page / Column		Page / Column		Page / Column		Page / Column	
1		1		1		1		1	
2		2		2		2		2	
3		3		3		3		3	
4		4		4		4		4	
5		5		5		5		5	
6		6		6		6		6	
7		7		7		7		7	
8		8		8		8		8	
9		9		9		9		9	
10		10		10		10		10	
11		11		11		11		11	
12		12		12		12		12	
13		13		13		13		13	
14		14		14		14		14	
15		15		15		15		15	
16		16		16		16		16	
17		17		17		17		17	
18		18		18		18		18	
19		19		19		19		19	
20		20		20		20		20	
21		21		21		21		21	
22		22		22		22		22	
23		23		23		23		23	
24		24		24		24		24	
25		25		25		25		25	
26		26		26		26		26	
27		27		27		27		27	
28		28		28		28		28	
29		29		29		29		29	
30		30		30		30		30	

Write the ANSWERS for your reusable work pages

Page / Column		Page / Column		Page / Column		Page / Column		Page / Column	
1		1		1		1		1	
2		2		2		2		2	
3		3		3		3		3	
4		4		4		4		4	
5		5		5		5		5	
6		6		6		6		6	
7		7		7		7		7	
8		8		8		8		8	
9		9		9		9		9	
10		10		10		10		10	
11		11		11		11		11	
12		12		12		12		12	
13		13		13		13		13	
14		14		14		14		14	
15		15		15		15		15	
16		16		16		16		16	
17		17		17		17		17	
18		18		18		18		18	
19		19		19		19		19	
20		20		20		20		20	
21		21		21		21		21	
22		22		22		22		22	
23		23		23		23		23	
24		24		24		24		24	
25		25		25		25		25	
26		26		26		26		26	
27		27		27		27		27	
28		28		28		28		28	
29		29		29		29		29	
30		30		30		30		30	

Write the ANSWERS for your reusable work pages

Page / Column		Page / Column		Page / Column		Page / Column		Page / Column	
1		1		1		1		1	
2		2		2		2		2	
3		3		3		3		3	
4		4		4		4		4	
5		5		5		5		5	
6		6		6		6		6	
7		7		7		7		7	
8		8		8		8		8	
9		9		9		9		9	
10		10		10		10		10	
11		11		11		11		11	
12		12		12		12		12	
13		13		13		13		13	
14		14		14		14		14	
15		15		15		15		15	
16		16		16		16		16	
17		17		17		17		17	
18		18		18		18		18	
19		19		19		19		19	
20		20		20		20		20	
21		21		21		21		21	
22		22		22		22		22	
23		23		23		23		23	
24		24		24		24		24	
25		25		25		25		25	
26		26		26		26		26	
27		27		27		27		27	
28		28		28		28		28	
29		29		29		29		29	
30		30		30		30		30	

Write the ANSWERS for your reusable work pages

Page / Column		Page / Column		Page / Column		Page / Column		Page / Column	
1		1		1		1		1	
2		2		2		2		2	
3		3		3		3		3	
4		4		4		4		4	
5		5		5		5		5	
6		6		6		6		6	
7		7		7		7		7	
8		8		8		8		8	
9		9		9		9		9	
10		10		10		10		10	
11		11		11		11		11	
12		12		12		12		12	
13		13		13		13		13	
14		14		14		14		14	
15		15		15		15		15	
16		16		16		16		16	
17		17		17		17		17	
18		18		18		18		18	
19		19		19		19		19	
20		20		20		20		20	
21		21		21		21		21	
22		22		22		22		22	
23		23		23		23		23	
24		24		24		24		24	
25		25		25		25		25	
26		26		26		26		26	
27		27		27		27		27	
28		28		28		28		28	
29		29		29		29		29	
30		30		30		30		30	

Write the ANSWERS for your reusable work pages

Page / Column		Page / Column		Page / Column		Page / Column		Page / Column	
1		1		1		1		1	
2		2		2		2		2	
3		3		3		3		3	
4		4		4		4		4	
5		5		5		5		5	
6		6		6		6		6	
7		7		7		7		7	
8		8		8		8		8	
9		9		9		9		9	
10		10		10		10		10	
11		11		11		11		11	
12		12		12		12		12	
13		13		13		13		13	
14		14		14		14		14	
15		15		15		15		15	
16		16		16		16		16	
17		17		17		17		17	
18		18		18		18		18	
19		19		19		19		19	
20		20		20		20		20	
21		21		21		21		21	
22		22		22		22		22	
23		23		23		23		23	
24		24		24		24		24	
25		25		25		25		25	
26		26		26		26		26	
27		27		27		27		27	
28		28		28		28		28	
29		29		29		29		29	
30		30		30		30		30	

Write the ANSWERS for your reusable work pages

Page / Column		Page / Column		Page / Column		Page / Column		Page / Column	
1		1		1		1		1	
2		2		2		2		2	
3		3		3		3		3	
4		4		4		4		4	
5		5		5		5		5	
6		6		6		6		6	
7		7		7		7		7	
8		8		8		8		8	
9		9		9		9		9	
10		10		10		10		10	
11		11		11		11		11	
12		12		12		12		12	
13		13		13		13		13	
14		14		14		14		14	
15		15		15		15		15	
16		16		16		16		16	
17		17		17		17		17	
18		18		18		18		18	
19		19		19		19		19	
20		20		20		20		20	
21		21		21		21		21	
22		22		22		22		22	
23		23		23		23		23	
24		24		24		24		24	
25		25		25		25		25	
26		26		26		26		26	
27		27		27		27		27	
28		28		28		28		28	
29		29		29		29		29	
30		30		30		30		30	

Write the ANSWERS for your reusable work pages

Page / Column		Page / Column		Page / Column		Page / Column		Page / Column	
1		1		1		1		1	
2		2		2		2		2	
3		3		3		3		3	
4		4		4		4		4	
5		5		5		5		5	
6		6		6		6		6	
7		7		7		7		7	
8		8		8		8		8	
9		9		9		9		9	
10		10		10		10		10	
11		11		11		11		11	
12		12		12		12		12	
13		13		13		13		13	
14		14		14		14		14	
15		15		15		15		15	
16		16		16		16		16	
17		17		17		17		17	
18		18		18		18		18	
19		19		19		19		19	
20		20		20		20		20	
21		21		21		21		21	
22		22		22		22		22	
23		23		23		23		23	
24		24		24		24		24	
25		25		25		25		25	
26		26		26		26		26	
27		27		27		27		27	
28		28		28		28		28	
29		29		29		29		29	
30		30		30		30		30	

Write the ANSWERS for your reusable work pages

Page / Column		Page / Column		Page / Column		Page / Column		Page / Column	
1		1		1		1		1	
2		2		2		2		2	
3		3		3		3		3	
4		4		4		4		4	
5		5		5		5		5	
6		6		6		6		6	
7		7		7		7		7	
8		8		8		8		8	
9		9		9		9		9	
10		10		10		10		10	
11		11		11		11		11	
12		12		12		12		12	
13		13		13		13		13	
14		14		14		14		14	
15		15		15		15		15	
16		16		16		16		16	
17		17		17		17		17	
18		18		18		18		18	
19		19		19		19		19	
20		20		20		20		20	
21		21		21		21		21	
22		22		22		22		22	
23		23		23		23		23	
24		24		24		24		24	
25		25		25		25		25	
26		26		26		26		26	
27		27		27		27		27	
28		28		28		28		28	
29		29		29		29		29	
30		30		30		30		30	

Write the ANSWERS for your reusable work pages

Page / Column		Page / Column		Page / Column		Page / Column		Page / Column	
1		1		1		1		1	
2		2		2		2		2	
3		3		3		3		3	
4		4		4		4		4	
5		5		5		5		5	
6		6		6		6		6	
7		7		7		7		7	
8		8		8		8		8	
9		9		9		9		9	
10		10		10		10		10	
11		11		11		11		11	
12		12		12		12		12	
13		13		13		13		13	
14		14		14		14		14	
15		15		15		15		15	
16		16		16		16		16	
17		17		17		17		17	
18		18		18		18		18	
19		19		19		19		19	
20		20		20		20		20	
21		21		21		21		21	
22		22		22		22		22	
23		23		23		23		23	
24		24		24		24		24	
25		25		25		25		25	
26		26		26		26		26	
27		27		27		27		27	
28		28		28		28		28	
29		29		29		29		29	
30		30		30		30		30	

Write the ANSWERS for your reusable work pages

Page / Column		Page / Column		Page / Column		Page / Column		Page / Column	
1		1		1		1		1	
2		2		2		2		2	
3		3		3		3		3	
4		4		4		4		4	
5		5		5		5		5	
6		6		6		6		6	
7		7		7		7		7	
8		8		8		8		8	
9		9		9		9		9	
10		10		10		10		10	
11		11		11		11		11	
12		12		12		12		12	
13		13		13		13		13	
14		14		14		14		14	
15		15		15		15		15	
16		16		16		16		16	
17		17		17		17		17	
18		18		18		18		18	
19		19		19		19		19	
20		20		20		20		20	
21		21		21		21		21	
22		22		22		22		22	
23		23		23		23		23	
24		24		24		24		24	
25		25		25		25		25	
26		26		26		26		26	
27		27		27		27		27	
28		28		28		28		28	
29		29		29		29		29	
30		30		30		30		30	

Write the ANSWERS for your reusable work pages

Page / Column		Page / Column		Page / Column		Page / Column		Page / Column	
1		1		1		1		1	
2		2		2		2		2	
3		3		3		3		3	
4		4		4		4		4	
5		5		5		5		5	
6		6		6		6		6	
7		7		7		7		7	
8		8		8		8		8	
9		9		9		9		9	
10		10		10		10		10	
11		11		11		11		11	
12		12		12		12		12	
13		13		13		13		13	
14		14		14		14		14	
15		15		15		15		15	
16		16		16		16		16	
17		17		17		17		17	
18		18		18		18		18	
19		19		19		19		19	
20		20		20		20		20	
21		21		21		21		21	
22		22		22		22		22	
23		23		23		23		23	
24		24		24		24		24	
25		25		25		25		25	
26		26		26		26		26	
27		27		27		27		27	
28		28		28		28		28	
29		29		29		29		29	
30		30		30		30		30	

www.ingramcontent.com/pod-product-compliance
Lightning Source LLC
Chambersburg PA
CBHW082347220526
45470CB00008B/2668